Arduino 电路与项目指南(影印版)

About the Author

The author of this book has been active for 16 years in research and development.

The main focus of his work for various major companies, such as Siemens and ABB, is Project Management in the areas of electronic development and physical technology.

Under his leadership and cooperation, serveral universities have received patents in various fields, from the electronics in environmental sensors to bio- and medical technology.

Thanks to his involvement in the fields of Sourcing Engineering and as Technology and Category Manager, he has detailed knowledge of the semiconductor market and production technology.

Besides his work as a specialist lecturer in physics and electrical engineering, he has published several articles and books on the topics of electronics, semiconductors and microcontrollers, and has created courses and tuition packages on these topics.

Arduino 电路与项目指南(影印版)

Günter Spanner 著

南京　东南大学出版社

图书在版编目(CIP)数据

Arduino 电路与项目指南：英文/(德)斯潘纳著．
影印本．—南京：东南大学出版社，2015.9
书名原文：Arduino：Circuits & Projects Guide
ISBN 978-7-5641-5948-1

Ⅰ.①A… Ⅱ.①斯… Ⅲ.①单片微型计算机-程序设计-英文 Ⅳ.①TP368.1

中国版本图书馆 CIP 数据核字(2015)第 178124 号

© 2013 by Elektor International Media BV

Reprint of the English Edition, jointly published by Elektor International Media BV and Southeast University Press, 2015. Authorized reprint of the original English edition, 2015 Elektor International Media BV, the owner of all rights to publish and sell the same.

All rights reserved including the rights of reproduction in whole or in part in any form.

英文原版由 Elektor International Media BV 出版 2013。

英文影印版由东南大学出版社出版 2015。此影印版的出版和销售得到出版权和销售权的所有者 —— Elektor International Media BV 的许可。

版权所有，未书书面许可，本书的任何部分和全部不得以任何形式重制。

Arduino 电路与项目指南(影印版)

出版发行：东南大学出版社
地　　址：南京四牌楼 2 号　　邮编：210096
出 版 人：江建中
网　　址：http://www.seupress.com
电子邮件：press@seupress.com
印　　刷：常州市武进第三印刷有限公司
开　　本：787 毫米×980 毫米　　16 开本
印　　张：16.25
字　　数：318 千字
版　　次：2015 年 9 月第 1 版
印　　次：2015 年 9 月第 1 次印刷
书　　号：ISBN 978-7-5641-5948-1
定　　价：56.00 元

本社图书若有印装质量问题，请直接与营销部联系．电话(传真)：025-83791830

Table of Contents

1   Introduction **9**
   1.1   The Arduino Project 10
   1.2   Book Structure and Target Audience 10

2   Structure and Function of the Hardware **11**
   2.1   Lots to Choose From: Arduino Versions 12
   2.2   Available for all Purposes: Shields 13
      2.2.1   Proto Shield 13
      2.2.2   Motor Shield 13
      2.2.3   Ethernet Shield 14
   2.3   Nothing Happens without Power: the Power Supply 15
   2.4   The Microcontroller: Heart of the Arduino 15
   2.5   The PC Connection: USB interface 16

3   Development Environment and Programming Basics **17**
   3.1   Integrated Development Environment (IDE) 17
   3.2   For Linux Users: Arduino IDE Under Ubuntu 23
   3.3   Simple Beginnings: The Warning Light 24
   3.4   General Structure of an Arduino Sketch 24
   3.5   Basic Elements of the C Programming Language for Arduino 27
   3.6   Program Structures 29
   3.7   Arduino-Specific Functions 29
   3.8   Arduino Punctuation: Syntax Elements 30
   3.9   Data Storage Elements: Variables 30
   3.10 The Art of Math: Operators 31
   3.11 To Be Clear: Logical Operators 32
   3.12 Unchanging Parameters: Constants 35
   3.13 Dealing with Related Data of the Same Type: Arrays 36
   3.14 Program Control Structures 36
   3.15 Controlling Timing 39
   3.16 Mathematical Functions 40
   3.17 Random numbers 40
   3.18 Extending the Options: Adding Libraries 41

4   Electronic Components and Low-Cost 'Freeduinos' **43**
   4.1   Breadboards: Simple and Effective without Soldering 43
   4.2   Prototyping Boards: Durable Construction without Chemicals 44
   4.3   Low-Cost 'Freeduinos' 45
   4.4   Arduino and Its Helpers: Basic Electronic Components 46
      4.4.1   USB Cable 46
      4.4.2   Resistors 46
      4.4.3   Capacitors 47
      4.4.4   Potentiometers 47
      4.4.5   LEDs 48
      4.4.6   RGB LEDs 48

TABLE OF CONTENTS

	4.4.7 Switches..	49
	4.4.8 Silicon Diodes..	49
	4.4.9 Transistors...	49
5	**Hello World**..	**51**
	5.1 Cut to the Chaser	51
	5.2 It Gets Brighter: Controlling Power LEDs	53
	5.3 POVino: Persistence-of-Vision Display	54
6	**Displays and Display Techniques**	**59**
	6.1 Bar Graph Display: The Classic for Measurement Applications	59
	6.2 Simple and Cheap: 7-Segment Displays...........................	59
	6.3 4-Digit, 7-Segment Displays: A Key Component for Instrumentation	62
	6.4 Mini Monitor for Signs and Graphics: The LED Dot Matrix.............	67
	6.5 Dot Matrix Display as a Two-Digit Digital Display	70
	6.6 Micro Learns to Write: Alphanumeric Display	72
	6.7 The LCD	76
7	**Measurement and Sensors**...	**81**
	7.1 Flexible and Easy to Read: An LED Voltmeter.................	81
	7.2 Volt / Ammeter: Precise Instrument for the Hobbyist's Lab	83
	7.3 Kiloohmmeter for Specific Applications...........................	87
	7.4 No More Faulty Electrolytic Capacitor Woes: The 'Elcaduino' Tester ..	89
	7.5 'Picofaraduino': Measuring Smaller Capacitances...........................	91
	7.6 'Transistino': Transistor Tester...	93
	7.7 A Simple NTC Thermometer	95
	7.8 Hot or Cold? Temperature Measurement Using the AD22100......	98
	7.9 Remote Thermometer...	99
	7.10 'Thermoduino': Precision Thermometer with 7-Segment Display......	100
	7.11 When Are We Most Comfortable? – The Hygrometer	105
	7.12 'Battduino': Capacity Measurement for Rechargeable Batteries........	107
	7.13 Optical Sensors: Important for More than Just Photography............	112
	7.14 Reflex Light for Geocaching	113
	7.15 For Professional Photographers: A Digital Light Meter.....	115
	7.16 Home 'Radar Station'': Distance Measurement Using Ultrasound.....	117
8	**Timers, Clocks and Interrupts**...	**123**
	8.1 Morning and Night Fun: Grand Prix Toothbrush Timer	123
	8.2 Practical and Accurate: Digital Clock with LED Display....	130
	8.3 Who's Faster? A Reaction Timer	135
	8.4 'Timerino': Universal Timer with a 7-Segment Display....	138
	8.5 Plug-in Timer to Make Life Easier	141
	8.6 Atomic Precision: The DCF77 Radio Clock......................	142
	8.7 Output of Time and Date to the Serial Interface	145
	8.8 Stand-Alone DCF77 Clock with LCD Display	146
9	**Interfaces**	**149**
	9.1 Universal and Simple: The I²C Interface	149
	9.2 When We Run Out of Pins: Port Expansion.....................	152
	9.4 Hexadecimal Debugger Display Using a 2-Digit, 7-Segment Display.	155
	9.5 LCD Control via I²C Using the PCF8574............................	158
	9.6 This Time, Fully Digital: The LM75 Thermometer...........	160
	9.7 Power-Saving: Real-Time Clock with Date Display	163

	9.8 After Including The IRremote Wireless, Practical, Quick: The IR Interface	167
	9.9 'Lampino': An IR-Controlled RGB Lamp	171
	9.10 Timely Luxury: An IR-Controlled Digital Clock	174
	9.11 Optimal for Microcontrollers: The PS/2 Interface	178
	9.12 Keyboard and Mouse as Universal Input Devices	178
	9.13 A Complete Microcomputer with LCD Monitor and Keyboard	181
10	**Sounds and Synthesizer**	**185**
	10.1 Simple Tones	185
	10.2 Transducers and Amplifiers	187
	10.3 Fast PWM Makes It Happen: Not Just tones, But Sound Waves	188
	10.4 Theremin: The Contactless Musical Instrument	195
	10.5 Audio Processing	197
	10.5.1 VCO: A Tunable Sine Wave Source	198
	10.5.2 Digital Signal Processing	200
	10.6 Sound Cloud: A Digital Synthesizer	203
11	**Digital Control Techniques**	**207**
	11.1 Control Types	208
	11.1.1 P Controller	208
	11.1.2 I Controller	209
	11.1.3 PI Controller	209
	11.1.4 PD Controller	209
	11.1.5 PID Controller	210
	11.2 Optimum Workstation Lighting: Digital Illumination Control	210
	11.3 A Classic of Control Theory: The Gravity Compensator	214
12	**Physical Computing**	**221**
	12.1 Servos Control the World	221
	12.2 'Photino': 2D Camera Swivel	225
	12.3 'Cranino': Mouse-Controlled Crane	
13	**Processing**	**231**
	13.1 Arduino and Processing: A Formidable Team	232
	13.2 Interaction with Processing: Data Logging, Trend Graphs, etc.	233
14	**The 'Living Room Box': Our Modular Concluding Project**	**241**
	14.1 Always Useful: A Clock	242
	14.2 Control from Afar: The IR Interface	243
	14.3 230 V Control for Hi-Fi systems, Televisions, Lamps, etc.	243
	14.4 Timers and Sensors as the Basis for Home Automation	243
	14.5 Indoor and Outdoor Thermometers	244
	14.6 No More Dry Air: A Hygrometer	244
	14.7 The Hardware	244
Bibliography		**249**
Listings		**251**
List of Figures		**253**
List of Tables		**257**
Index		**259**

1
Introduction

There are two main reasons for Arduino's success. The first is the complete processor board, which has significantly eased entry into the microcontroller hardware arena. Typical beginner problems, such as bad power supplies, problems with setting the configuration parameters (fuse bits) and crystals that won't oscillate due to bad load capacitances are all foreign to the Arduino world. The board is simply connected to a PC's USB port, and off we go. People young and old who've never been involved in electronics won't have any difficulty.

The second success factor is the associated programming interface, which is provided as free-of-charge open source software. The second success factor is the associated programming interface, which is provided as free-of-charge open source software. In addition, installation is quick and easy, so the environment is usable immediately. Simple introductory examples encourage rapid progress. The selection of complicated parameters is not required, and the first example programs may be opened and run within minutes.

Furthermore, the Arduino is backed up with a wealth of software libraries, and the number of which grows daily, often presenting beginners with their first problem: after the introduction of simple examples, the way forward is not clear. This is often due to the lack of detailed descriptions and explanations. The number of projects on the internet, which are explained to a greater or lesser extent, is more likely to cause confusion. Because these applications are designed by a number of different people, each with their own goals in mind, there is no common thread connecting them.

That's where this book comes in. Projects are introduced systematically, each introducing a different theme. A practical hands-on approach is employed alongside the necessary theoretical foundations and, in a similar vein, important concepts such as A/D conversion, timers and interrupts are presented using practical projects. There are running lights, fully-functional voltmeters, precise digital thermometers, clocks of all kinds, reaction timers and a mouse-controlled robot crane. Along the way, the reader will gain an understanding of the associated controller techniques and pick them up fully — in the truest sense of the word.

The practical projects presented herein will not be relegated to the status of mere 'laboratory prototypes'. By means of appropriate hints and notes, other practical devices will arise, which may be used for home, hobby and work. The projects are always implemented using easy-to-find and inexpensive components.

In the final chapter, an Arduino-based 'Living Room Box' is presented. It is designed modularly and may be adapted to individual requirements. The knowledge gleaned throughout the book is used practically to produce a very useful, yet unusual, device.

1 INTRODUCTION

1.1 The Arduino Project

The Arduino concept came to be in 2005 at the Institute for Interactive Design in Ivrea, Italy. The search for a low-cost microcontroller system for design students led to a handy printed circuit board that contained all of the necessary electronic building blocks. The main objective was the development of an inexpensive microcontroller board that could be utilized quickly and simply by art and design students who had no previous knowledge of programming or electronics.

The first version of the Arduino hardware consisted of a kit that could be soldered together easily, and it sold out quickly. Newer versions followed in rapid succession. Designers and artists from other regions took up the idea and the Arduino principle spread, first in Italy, then into Europe, and eventually to the rest of the world.

Interest grew rapidly in application areas outside of art academies and design schools. The concept of a simple and low-cost hardware platform with a freely available, easy-to-learn programming language was quickly embraced by hobbyists. Finally, schools and universities in the scientific and technical fields recognized the enormous potential of the Arduino idea. New hardware versions, as well as plug-in expansion boards, or 'shields', arose, and Arduino application areas were limited only by users' imaginations.

Meanwhile, the number of Arduinos shipped has exceeded the 100,000 mark by far. If DIY versions and clones are considered as well, the actual figure may exceed 1 million. It is, therefore, no exaggeration to claim that the Arduino microcontroller board is the most successful of all time.

1.2 Book Structure and Target Audience

This book is aimed at those who've already had some basic experience in the electronics field. The typical high school lessons on circuits, Ohm's law, etc., are fully sufficient. In Chapter 4, the basic functions and characteristics of the major components are explained. In due course, more elaborate projects are discussed, so that even students and teachers who come with some pre-existing technical knowledge will encounter new challenges.

The projects are grouped into individual themes. However, care was taken to place the less technically challenging material at the beginning of the book. For this reason, beginners will find it helpful to go through the book's chapters in order, even if not every single project is constructed.

2 Structure and Function of the Hardware

In its original form, the Arduino board was fitted with an ATmega8 microcontroller and a simple RS-232 level converter. In this way, it could be programmed directly from a PC's serial port. Over time, a variety of updates and versions has evolved from this prototype. Since fewer and fewer PCs came equipped with this legacy interface over the years, one of the first major steps in Arduino development was the addition of a USB-to-RS-232 converter. At first, an FTDI chip was used. Later, on the Arduino UNO board, this relatively expensive chip was replaced by an ATmega8U2. Using suitable firmware, this chip is able to manage USB-to-RS-232 protocol conversion.

Figure 2.1:
Arduino Uno SMD

The original ATmega8 processor was also replaced by the more powerful ATmega168 and ATmega328 variants. In parallel with the main line of development, several more-or-less compatible offshoots have emerged over the years. For example, there are rugged versions in which all ports are specially protected against overvoltage and short circuits, and special versions with custom form factors and pin spacings.

2 STRUCTURE AND FUNCTION OF THE HARDWARE

2.1 Lots to Choose From: Arduino Versions

Besides the basic Arduino reference design, there are many other versions in various sizes and form factors. Examples include the Arduino MEGA, equipped with an ATmega1280 or ATmega2560. These processors have a much wider range of functions and a significantly greater number of available pins. Due to the greater pin count, the boards are roughly double the size of the classic Arduino.

On the other end, if you want to build small, compact devices, you can go down to the Arduino Mini or Arduino Nano devices, which are each roughly the size of a postage stamp. Instead of connector sockets, they have only solder pads, onto which cables may be directly soldered. Alternatively, pin headers may be attached, so that these compact boards may be plugged directly into breadboards or IC sockets.

The LilyPad is also of particular note, characterized by its special form factor — not a regular rectangular circuit board but a circular disc. The main application area for this interesting Arduino variant is in the field of 'wearable computing', in which the LilyPad is sewed into articles of clothing and connected to LEDs, sensors or actuators via thin wires or conductive threads.

Finally, the 'pin spacing dilemma' should be mentioned. Unfortunately, in the development of the classic Arduino, not all of its connectors were aligned to the common 1/10" (2.54 mm) grid spacing. In fact, the top two female headers have a spacing between them of only 1/20", so, unfortunately, standard prototyping boards cannot be connected directly to all of the connectors using pin headers. Here, one always has to resort to work-around solutions.

Whether this layout was deliberately chosen to prevent the easy development of plug-in replica shields or whether it was a simple oversight has stirred up lengthy discussions on internet forums.

Meanwhile, other Arduino variants that followed have also been manufactured with this aesthetic flaw, as compatibility with the many preexisting shields dictates the necessity of maintaining the original pin spacing.

The following figure shows the Arduino pin assignments. The names of the embedded microcontroller's ports are also shown. In this way, we make a clear connection between the Arduino world and the world of the microcontroller professional. Initially, a simple understanding of Arduino terminology is sufficient. Later on, however, when more advanced applications are realized, the Arduino pin assignment map will prove very useful.

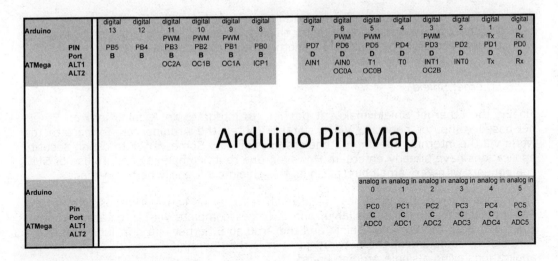

Figure 2.2: Arduino Pinout

2.2 Available for all Purposes: Shields

Plug-in modules for the Arduino are called shields, and they greatly expand the board's functionality. There are over 100 different types of shields available, ranging in functionality from simple prototyping to Ethernet connection, XBee wireless, motor drivers, to color graphics display shields. What follows is a brief description of the most important and best-known shields.

2.2.1 Proto Shield

There are many different varieties of this shield. In the simplest case, a proto shield consists of a stripboard with two soldered pin headers. The only difference from an ordinary stripboard is that the second digital connector is spaced appropriately from the first, specifically for the Arduino (see 'pin-spacing-dilemma' in the previous section), and the module can thus be connected without any mechanical obstacles.

Other variations include solderless breadboards, with which simple prototypes may be quickly and robustly created.

It isn't very difficult to create suitable DIY shields, so one should decide whether the situation merits the purchase of a more expensive proto shield.

2.2.2 Motor Shield

These have various power drivers to enable the control of different types of electric motors. Depending on the version, DC and stepper motors may be driven. These boards usually have screw terminals that enable the connection of the thicker motor wires securely.

2 STRUCTURE AND FUNCTION OF THE HARDWARE

Because model servo motors already contain very practical and affordable motors, this book dispenses with a discussion of motor shields. However, should larger robot systems with stronger actuators be considered, motor shields may be quite useful.

2.2.3 Ethernet Shield

Finally, the Ethernet Shield makes it possible to integrate an Arduino into an Ethernet-based computer network. Via a suitable router, the Arduino can connect to the world via the internet, opening up endless possibilities. Some very interesting security applications have already emerged. However, due to its complexity and its use of SMD components with very small pin spacings, this shield can't easily be home-made.

On the other hand, the Ethernet Shield might already be considered obsolete, as the Arduino Ethernet board is available, and it comes complete with integrated Ethernet controller. Both the Ethernet Shield and the Arduino Ethernet board include a microSD card slot to complement the network jack. It is thus possible to create complete web applications using a single Arduino board.

Figure 2.3:
Typical DIY shield

In summary, it may be said that, while shields offer quick and practical solutions for many applications, they are usually a little expensive, so, before buying a shield, it's always advisable to consider whether building one's own would be more beneficial.

2.3 Nothing Happens without Power: the Power Supply

To power the Arduino, there are two available options:

- power from the USB port
- an external supply via the separate power socket

Since the original version of the Arduino had only a classic RS-232 interface, it required an external power supply. That's why the Arduino still has a power socket and a voltage regulator. With the USB version, this external power supply may be unnecessary, as the board is directly powerable via the USB interface. However, should deployment in the field be required without an available PC connection, the Arduino can be powered using a simple power adapter with a USB socket.

Most USB-powered Arduino devices also have a separate power connector. At minimum, this has the advantage of the device being powered using either batteries or cheap power adapters. Thanks to the onboard regulator, a wide range of external voltages is suitable. If only a few external components are driven by the board, then a DC supply of between 6 and 20 V suffices. The optimum range is, however, at

$$V_{min} = 7 \text{ V and } V_{max} = 12 \text{ V}$$

to ensure a greater margin of safety for unexpected voltage fluctuations. In principle, even higher input voltages are possible, although damage to the voltage regulator on the Arduino board cannot be ruled out at higher currents. Thus, the external supply voltage need not be especially stabilized, but, for safe operation of the controller, the supply should be able to deliver a current of at least 300 mA.

When connecting to an external power supply, note that the outer conductor is connected to the negative of the power supply and the inner to the positive.

Figure 2.4: Power Supply Connector Polarity

Note
When powering the board from an external supply, don't forget to move the jumper (if applicable to the respective board) from the 'USB' to the 'EXT' position.

2.4 The Microcontroller: Heart of the Arduino

The most important component on the Arduino board is, of course, the microcontroller. In the classic Arduino, one of two processors was used:

- ATmega168P
- ATmega328P

2 STRUCTURE AND FUNCTION OF THE HARDWARE

These two differ only in the amount of available memory. Key features of the two are listed below:

ATmega168P	ATmega328P	
16	32	KB flash memory
0.5	1	KB EEPROM
1	2	KB SRAM
2	2	8-bit timers/counters
1	1	16-bit timer/counter capture
6	6	PWM channels
6	6	10-bit A/D channels
23	23	programmable I/Os

Table 1: *Arduino ATmega168P and ATmega328 Microcontroller Features*

In addition, both controller versions offer the following:

- programmable serial USART
- Master/Slave SPI serial interface
- 2-wire serial interface (Philips I²C-compatible)
- watchdog timer
- on-chip analog comparator
- internal calibrated oscillator
- 1.8 – 5.5 V operating voltage
- 0 – 20 MHz clock frequency at 1.8 – 5.5 V
- Power consumption (@1 MHz, 1.8 V, 25 °C):
  Active Mode: 0.2 mA
  Power-down Mode: 0.1 µA
  Power-save Mode: 0.75 µA (incl. 32 kHz RTC)

2.5 The PC Connection: USB interface

In the current version, communication with the PC usually takes place exclusively via USB. The interface serves not only as the programming interface, but is also used to exchange data between the controller and the computer during program execution. For example, an Arduino could be used to send temperature readings to a PC periodically, where the readings may be saved and processed in a spreadsheet application such as Microsoft Excel or OpenOffice.

3 Development Environment and Programming Basics

The Arduino board is programmed via a special, user-friendly integrated development environment (IDE). The IDE's biggest advantage in comparison with the classic tool chain is that it's very intuitive. In addition to the Arduino board itself, this special simplified IDE is certainly one of the main factors in the success of the Arduino concept.

3.1 Integrated Development Environment (IDE)

Installation of the Integrated Development Environment (IDE) is not particularly tedious, and, subsequent to this, the first programs can be loaded onto the microcontroller immediately after starting the IDE and selecting the correct board and serial port.

The current version of the IDE is always available for free at

arduino.cc/en/Main/Software

There are versions available for the most common desktop operating systems (Windows, Mac OS and Linux).

The complete package is contained in a compressed Zip file and may be extracted to the folder of choice. Note that the latest version comes as a normal executable installer.

Note
> At the time of writing the current version of the Arduino IDE is 1.0.5. The previous versions were all indicated using four-digit numbers (0001 to 0023). Unfortunately, in the switch to Version 1.0, many changes were implemented that greatly reduced compatibility with preexisting sketches and libraries.
>
> Some of the programs in this book must therefore be loaded and worked on in Arduino 0022. These programs can be identified by their `.pde` extension.
>
> Programs for Version 1.0 and up have the extension `.ino`.
>
> This incompatibility has been the source of some discord among Arduino users. However, one will certainly have to live with this transitional solution for some time, at least until the most important libraries and sketches have been ported to the new version.

3 DEVELOPMENT ENVIRONMENT AND PROGRAMMING BASICS

Figure 3.1: *Typical Arduino Directory*

The `Arduino-1.0` directory contains all of the software necessary for programming the board.

In addition, several example programs are available in the `examples` subdirectory.

The `libraries` subdirectory contains useful program libraries, which may be used to control

> stepper motors and model servos
> LCDs
> external components such as switches, sensors and EEPROMs
> etc.

In addition, the Arduino's capabilities may be supplemented using additional custom libraries. More on this topic is found in Section 3.18.

After unpacking the Zip archive, the IDE is started using the `arduino.exe` program file in the `./arduino-1.x.x` directory. (For newer versions that make use of an installer, a shortcut will be added to the Start menu.)

Figure 3.2: *Arduino Shortcut for Starting the IDE*

After successfully starting the program, the following splash screen appears with the usual information about the authors of the program:

INTEGRATED DEVELOPMENT ENVIRONMENT (IDE) 3.1

Figure 3.3:
Arduino IDE 1.0.5 Splash Screen

Then, the IDE window is displayed:

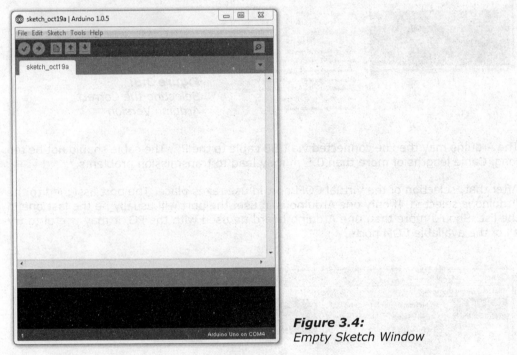

Figure 3.4:
Empty Sketch Window

If no program has been loaded, or the new file has not yet been saved under a different name, a default name is created:

```
sketch_mmmddx
```

This consists of the word `sketch`, an underscore, and an abbreviation for the current month and current day, as well as a letter. The letters `a`, `b`, `c`, etc. stand for the 1st, 2nd, 3rd, etc. new sketch of the current day.

3 DEVELOPMENT ENVIRONMENT AND PROGRAMMING BASICS

In the next step, the correct Arduino version should be selected.

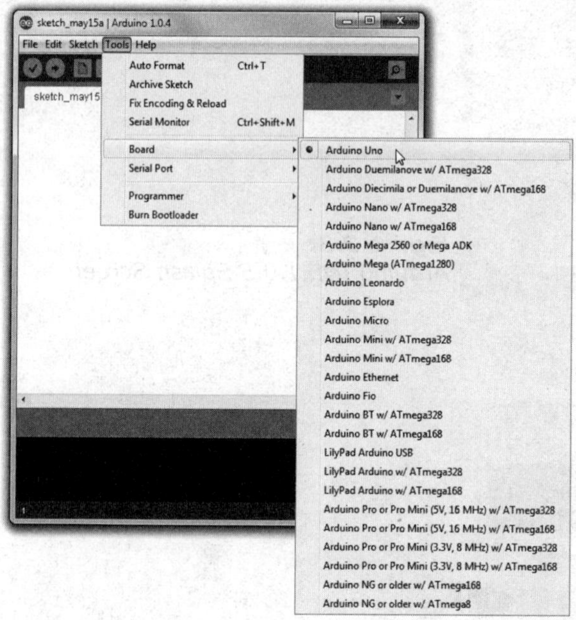

Figure 3.5:
Selecting the Correct Arduino Version

The Arduino may then be connected via USB cable to the PC. The cable should not be too long. Cable lengths of more than 0.5 m may lead to transmission problems.

After that, selection of the virtual COM port in use takes place. The port assigned to the Arduino is selected. If only one Arduino is in use, the port will usually be the last one in the list. Should more than one Arduino board be used with the PC, it may useful to try all of the available COM ports.

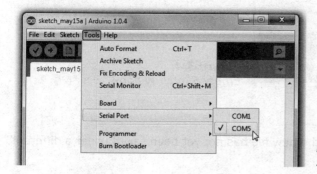

Figure 3.6:
Selection of the Virtual COM Port

INTEGRATED DEVELOPMENT ENVIRONMENT (IDE) 3.1

Below the menu bar, at the top of the IDE, you will find the following icons:

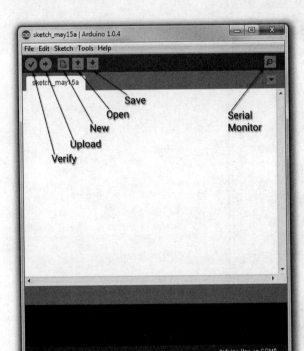

Figure 3.7:
IDE Icons

These have the following meanings:

- **Verify**: Begins compilation of the program. Should the compiler detect errors in the code, these will be displayed in the black information area at the bottom of the window.

- **Upload**: Compiles the sketch then uploads it to the Arduino board.

- **New**: Begins a new sketch.

- **Open**: Opens an existing sketch from the current working directory.

- **Save**: Saves the current sketch to disk.

- **Serial Monitor**: Opens the Serial Monitor window, which is used to monitor communication between the Arduino and the PC during program execution.

We may now load our first program into the IDE using the ⬆ *Open* icon REPLACE WITH open_icon.png. We'll choose this program:

        ..\Arduino\examples\01.Basics\Blink\Blink.ino

3 DEVELOPMENT ENVIRONMENT AND PROGRAMMING BASICS

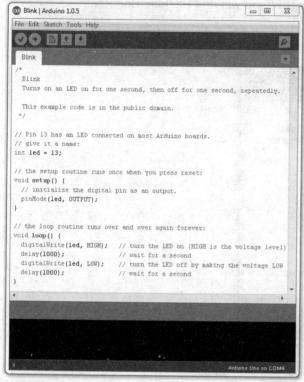

Figure 3.8:
Blink *Sketch Loaded in the IDE*

Using the ⬆ *Upload* button REPLACE WITH upload_button.png, the program is transferred to the microcontroller's memory.

The Rx/Tx LEDs flash briefly at irregular intervals during program upload. This is a useful indication of traffic on the serial port.

At the same time, the message

Uploading to I/O Board...

appears in the IDE.

After the transfer is complete, the message

    Done uploading.

appears, the Rx and Tx LEDs stop flashing, and LED 13 starts blinking regularly, as per the program.

Figure 3.9:
LED 13 in action

After data transfer to the Arduino is complete, an automatic reset takes place and the *Blink* program starts running.

3.2 For Linux Users: Arduino IDE Under Ubuntu

The Arduino IDE is also available for users of the Linux operating system. In current versions of Ubuntu, installation is very simple, as the IDE is available in the Ubuntu Apps Directory.

The Linux IDE is almost identical to the Windows version. The only real disadvantage to the Linux version is that releases of the latest versions of the IDE usually come out a little later.

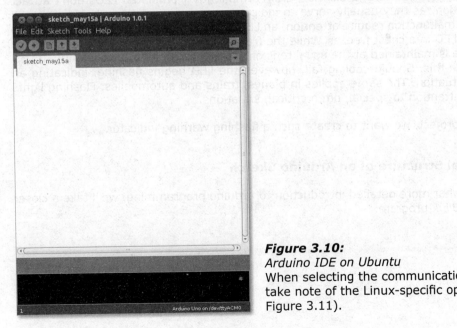

Figure 3.10:
Arduino IDE on Ubuntu
When selecting the communication port, take note of the Linux-specific options (see Figure 3.11).

3 DEVELOPMENT ENVIRONMENT AND PROGRAMMING BASICS

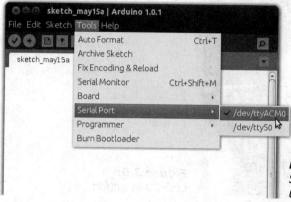

Figure 3.11:
Selection of the Serial Interface under Linux

Program editing and uploading takes place in the same manner as under Windows. The availability of the IDE to the Linux community is no small advantage to Arduino, as Linux users are often prevalent in exactly this area of microcontroller programming and embedded development.

3.3 Simple Beginnings: The Warning Light

LEDs are prevalent in all areas of life these days. In home entertainment devices, they indicate operating status. In automobiles, they have long ago replaced the incandescent oil and alternator warning lamps. LEDs indicate the status of everything from laptops and monitors to freezers and dishwashers. Permanently illuminated LEDs don't attract much attention, as they usually serve to indicate a normal condition. Should a special situation or malfunction require attention, an LED is usually flashed. A typical example is the control LED in a chest freezer. While the freezer is working correctly and its interior temperature is maintained at the usual tens of degrees below zero, a constantly lit LED may indicate this. Should cooling fail, however, the LED begins flashing, indicating an abnormal situation. The same applies in planes, trains and automobiles. Flashing lights draw clear attention to special, often critical, situations.

In our first project, we want to create such a flashing warning indicator.

3.4 General Structure of an Arduino Sketch

For a somewhat more detailed introduction to Arduino programming, we'll take a closer look at the *Blink* program.

SIMPLE BEGINNINGS: THE WARNING LIGHT   3.3

```
/*
  Blink
  Turns on an LED on for one second, then off for one second, repeatedly.

  This example code is in the public domain.
*/
// Pin 13 has an LED connected on most Arduino boards.
// give it a name:
int led = 13;

// the setup routine runs once when you press reset:
void setup() {
  // initialize the digital pin as an output.
  pinMode(led, OUTPUT);
}

// the loop routine runs over and over again forever:
void loop() {
  digitalWrite(led, HIGH);   // turn the LED on (HIGH is the voltage level)
  delay(1000);               // wait for a second
  digitalWrite(led, LOW);    // turn the LED off by making the voltage LOW
  delay(1000);               // wait for a second
}
```

Listing 3-1: *Blink.ino*

This program line is a comment:

```
// Pin 13 has an LED connected on most Arduino boards.
```

Comment lines begin with a double slash (//). Characters following the double slash are ignored by the compiler. They serve only to make the code more readable.

The use of comments should, nonetheless, be encouraged. Experience shows that programs lacking good comments, even after a short time, cannot be understood without extensive scrutiny, even by the original programmer.

Comments should be brief, but they should make the function of a program block clear. Not every program line should be commented; it is rather more useful to explain the purpose of specific logical blocks in plain language.

The first active line

```
int led = 13;
```

sets Pin 13 as the LED pin. On the Arduino Uno board, this pin is hard-wired to an LED via a resistor. Pin 13 is therefore often used for test purposes.

With this definition in place, `ledPin` may be used in the sketch wherever Pin 13 is meant. Definitions such as this make programming much easier. It's not practical to memorize the functions of all of the pin numbers and their assignments, especially in larger applications,

25

3 DEVELOPMENT ENVIRONMENT AND PROGRAMMING BASICS

so, meaningful, plain-language tokens simplify the programming task significantly.

The next active line

```
void setup() {
```

is a speciality of the *Processing* programming environment, upon which the Arduino IDE is based (see Chapter 13). `setup()` is an initialization function that establishes, among other things, what purposes the individual Arduino pins serve.

In our case, `led`, a variable created above, is used to assign Pin 13 as an output:

```
pinMode(led, OUTPUT);
```

The `setup()` function is always required in an Arduino sketch, even if this function is empty. When `setup()` is called, internal processes are triggered that start the Arduino working.

The start of the main program is

```
void loop() {
```

In microcontrollers, the main program is almost always an endless loop. Because controllers usually don't have their own operating systems, the main program loop should not end. Should the 'end' of a program be reached, the controller would simply stop doing anything useful and would only be usable again after a reset. `loop()` enforces this endless loop, ensuring that the processor continues to work as long as power is supplied to the device.

With

```
digitalWrite(ledPin, HIGH);  // LED on
```

Pin 13 is set to a high potential. This means that the anode of the LED is connected to 5 volts, causing it to light up.

The duration for which the LED lights up is set using

```
delay(1000);  // wait 1 sec = 1000 ms
```

This function instructs the program to wait for 1 second (1,000 ms). Finally, with

```
digitalWrite(ledPin, LOW);   // LED off
```

the LED is turned off again and this is followed by another 1-second pause.

It then starts all over, as the main program is a continuous loop. The LED flashing sequence occupies a period of 2 seconds, corresponding to a frequency of $f = 1/(2\text{ s}) = 0.5$ Hz.

3.5 Basic Elements of the C Programming Language for Arduino

The most important basics of Arduino programming are explained here. Basically, the Arduino is programmed in standard C. The AVR-GCC standard C compiler, which has gained widespread adoption in microcontroller programming, is integrated into the IDE.

It is possible to borrow elements from standard C, and some of the projects in this book do exactly that. This may be advantageous if the Arduino-specific routines are too slow or specialized.

But why is a C version especially for the Arduino necessary? The answer is simple: pure C programs tend to scare beginners off. Compared to other languages encountered by beginners, such as BASIC (or BASCOM for microcontrollers), pure C code looks very cryptic.

This is exactly where the Arduino ideology comes into play. Several often-needed functions are predefined, e.g. `digitalWrite()`, greatly simplifying the beginner's journey. One obvious advantage is that the new user need not learn a specific language purely for beginners. Another is that, after gaining some experience, the user may easily switch to pure C programming. This is one of the greatest advantages of the Arduino IDE over other beginner languages such as BASCOM.

Since there is more than enough literature available about programming in C, only the important fundamentals will be explained here. Particular emphasis is placed on the language elements especially relevant to the Arduino. For more comprehensive information, see the [1] in the bibliography.

The following information is intended for newcomers to the platform. For even more information, the IDE's *Help* menu may be consulted. For context-specific help, simply highlight the instruction of interest, then select *Find in Reference* from the *Help* menu. The reference for the selected instruction will appear.

3 DEVELOPMENT ENVIRONMENT AND PROGRAMMING BASICS

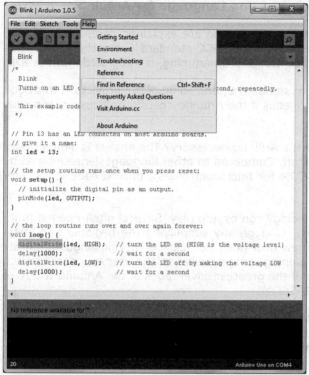

Figure 3.12:
Online Help

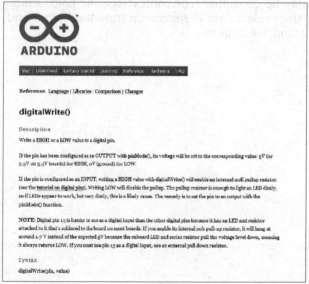

Figure 3.13:
Excerpt of the Help Text for the `digitalWrite()` Function

3.6 Program Structures

An Arduino sketch always consists of two main blocks:

> `void setup() {}`

and

> `void loop() {}`

The `setup()` function sets up all of the microcontroller's standard functionality. For this reason, it is required in all programs. The first instructions are usually found here. In `setup()`, pins are often preconfigured as inputs or outputs, e.g. `pinMode(ledPin, OUTPUT);`, etc.

The main program's code is contained within the `loop()` function. All of the commands in the function are executed in sequence until the controller is reset or powered down.

3.7 Arduino-Specific Functions

These functions are made available to simplify the entry into the world of microcontroller technology. Some of them are already familiar from the *Blink* program. The most important ones are described below.

`pinMode`(*pin, mode*)

> Here, the pin is configured as either an output or as an input (*mode* = `INPUT` or *mode* = `OUTPUT`).

`digitalWrite`(*pin, value*)

> This statement sets *pin* to either high potential or low potential (*value* = `HIGH` or *value* = `LOW`).

`digitalRead`(*pin*)

> reads a pin's level and returns either `HIGH` or `LOW`.

`analogWrite`(*pin, value*)

> sets the specified PWM output to the analog value *value*, where *value* is between `0` and `255`.

`analogRead`(*pin*)

> reads and returns the analog voltage applied to a pin (returning 10-bit values in the range 0 – 1023).

3.8 Arduino Punctuation: Syntax Elements

Like any programming language, Arduino C requires some syntax elements to indicate comments or to define command blocks.

The most important syntax elements are:

```
;    semicolon
```

The semicolon separates instructions. It usually appears at the end of a line, but may also be used within a line to separate short instructions.

```
{}   braces
```

Instruction blocks are enclosed in braces

```
//   single-line comment
```

Single-line comments must be preceded by a double slash. The comment terminates at the end of the line.

```
/* ... */    multi-line comment
```

When a comment needs to span several lines, the symbol combination /* is used at the beginning of the comment, and */ at the end.

Example:

```
/* This comment spans
several lines
*/
```

These characters are also often used to comment out parts of the program, perhaps for test purposes. Should the section not be required, it may be switched out using a multi-line comment.

3.9 Data Storage Elements: Variables

Variables are the most important elements in any programming language. A variable is defined using

```
int variableName = value;
```

If *value* is omitted from the definition, the variable is initialized to 0.

The statement

```
int addend = 5;
```

assigns a value of 5 to the variable addend.

Variable values change during the course of a program. Often, they are used for calculating results:

    sum = addend + augend;

adds `augend` to `addend` and stores the sum in `sum`.

Variables always consume a certain amount of memory, and, since memory is generally a scarce commodity in microcontrollers, different variable types have been provided in the Arduino programming language. Depending on their range of values, these variables will occupy a greater or lesser amount of memory. Table 2 shows an overview of the different variable types.

Type	Bytes	Range of values
byte	1	0 – 255
int	2	-32,768 – 32,767
word	2	0 – 65,535
long	4	-2,147,483,648 – 2,147,483,647
unsigned long	4	0 – 4,294,967,295
float	4	-3.4028235E+38 – 3.4028235E+38

Table 2: *The Most Important Variable Types*

In general, one should always choose the variable type that best suits the requirements, yet occupies the least amount of memory.

3.10 The Art of Math: Operators

In order to process variables, mathematical operators are needed. Operators are the basic elements of data processing.

Arithmetic Operators

The C arithmetic operators have their usual mathematical meanings:

    +   addition
    -    subtraction
    *   multiplication
    /   division
    %   modulus, i.e. the remainder of a division

3 DEVELOPMENT ENVIRONMENT AND PROGRAMMING BASICS

Comparison Operators

Here, too, the usual mathematical meanings apply, the only difference being that the symbols are represented using standard ASCII characters:

==	equal to
!=	not equal to
<	less than
>	greater than
<=	less than or equal to
>=	greater than or equal to

Assignment Operator

=	assignment

Note:

A common error is that of confusing the assignment operator `=` with the comparison operator `==`. The former is used to assign a certain value to a variable, e.g. `divisor = 10` assigns the value of 10 to the variable `divisor`, while the latter compares two values, e.g. `if (dividend == 1)` compares the value of the variable `dividend` with 1. One way of preventing this error in a comparison operator is to always place the number on the left-hand-side, e.g. `if (1 == dividend)`. That way, if one accidentally uses the assignment operator instead of the comparison operator, the compiler will generate an error message and the sketch will not compile until this is corrected. For more on comparison operators, see Section 3.14.

3.11 To Be Clear: Logical Operators

Logical operations are performed using the following operators:

Bitwise Operators

&	AND
|	OR
^	XOR
~	bitwise inversion
<<	left shift
>>	right shift

These operators have their usual binary meanings. The AND operator returns a 1 only when both arguments are 1.

AND:

```
0 & 0 == 0
0 & 1 == 0
1 & 0 == 0
1 & 1 == 1
```

The operators are applicable to integers other than 0 and 1. In those cases, all of the binary bits of the number are compared individually:

    0b00001010 & 0b00000010 == 0b00000010

Or, put it another way, and the operation is easier to see:

    0b00001010 &
    0b00000010 ==

    0b00000010

Only when there is a 1 in the same column of both arguments, will the result in that column be 1.

The same applies for the other logical operators:

OR:

    0 | 0 == 0
    0 | 1 == 1
    1 | 0 == 1
    1 | 1 == 1

XOR:

    0 ^ 0 == 0
    0 ^ 1 == 1
    1 ^ 0 == 1
    1 ^ 1 == 0

Inversion:

    ~0 == 1
    ~1 == 0

Shift operators are also often-used. They move bit patterns either to the left or to the right. They are easiest to understand in binary form. Shifting bits in a binary integer by n digits to the left is equivalent to multiplying the integer by 2^n. As an example:

    (2 << 3): The integer 2 is shifted by 3 places to the left.

In binary notation, this means:

    0b0000010 **becomes** 0b0010000

or, in decimal notation:

2 shifted left 3 positions is 2 x 2^3, or 16:

0b00000010 == 2_{dec}

0b00010000 == 16_{dec}

A great advantage of the shift operators is that they are executed very quickly, as the C compiler converts the shift operation directly into the corresponding bit operation. So, when one needs the fastest possible execution speed in a program block, one should consider whether complex algorithms could possibly be replaced by simple bit-shift operations.

Boolean Arithmetic

Similar to the bitwise operators, there are the Boolean algebra operators. These are not processed in a bitwise fashion. Rather, only two values, TRUE and FALSE, are considered.

To distinguish these from the bitwise operators, Boolean arithmetic operators use double characters:

    &&   AND

    || OR

Boolean inversion, however, uses its own character:

    !   NOT

In these examples, you get the following example results:

```
int a = 3;
(a > 0) && (a < 10); // returns TRUE
```

Similarly, the following applies:

```
int a = 12;
(a > 0) && (a < 10); // returns FALSE
```

A typical use of the Boolean AND operator is to query whether a value is within a given range:

```
if ((a >= 0) && (a < 10)) {
b = b + 1;
}
```

In this case, b is only incremented when a is greater than 0 and less than 10.

Compound Operators

The compound operators can make the C programming language look somewhat cryptic, but they are very easy to explain, and they save on a lot of typing.

++	increment, i.e., instead of x = x + 1, use x++
--	decrement, i.e., instead of x = x - 1, use x--
+=	compound addition, i.e., instead of x = x + y, use x += y
-=	compound subtraction, i.e., instead of x = x - y, use x -= y
*=	compound multiplication, i.e., instead of x = x * y, use x *= y
/=	compound division, i.e., instead of x = x / y, use x /= y

3.12 Unchanging Parameters: Constants

Predefined values, or constants, make programs more readable. In the Arduino IDE, a few often-used constants are predefined:

TRUE, FALSE

These two Boolean constants, mentioned above, are used to define logical values:

> FALSE is defined as 0 (zero)
> TRUE is anything not equal to 0

HIGH, LOW

These two constants represent pin levels as either high potential or low potential, and are used to read from and write to the digital pins.

> HIGH means a logic level of '1', i.e. ON or 5 volts
> LOW means a logic level of '0', i.e. OFF or 0 volts

INPUT, OUTPUT

These constants are provided for the `pinMode()` function to represent whether a pin is to be used as an input or as an output.

In addition to these predefined values, user-defined constants may also be set. Using `#define`, a constant is assigned a desired value. A typical example would be defining a constant named PI to represent a value of 3.14159, or to establish a port's state using the constants ON or OFF.

```
// program constants
#define PI 3.14159
#define OFF 0
#define ON 1
```

It is customary to use all uppercase characters to name constants. This way, they can easily be distinguished from variables at any point in a program.

3.13 Dealing with Related Data of the Same Type: Arrays

Arrays systematically represent a collection of variables of the same type. Each individual value in an array may be referenced using an index number.

All of the following statements create a fixed-size array:

```
int myHappyNumbers[5];
int outputPins[] = {1, 1, 2, 3, 5};
int optoSensorValues[9] = {1, 1, 4, -3, 5, -6, 1, 2, 4};
char standardGreeting[6] = "Hello";
```

With these respective array definitions, arrays of 5, 5, 9 and 6 elements are created, respectively.

Referencing the array element is done using the relevant index, e.g.:

```
outputPins[0] = 1;
x = optoSensorValues[4];
```

Important:

>Array indices begin at 0, i.e. the first element of the array has an index of 0.

This means that, for an array of 9 values, 8 would be the last indexed value. If index number 9 is retrieved from an array of 9 elements, an unknown value will be returned!

```
int myHappyNumbers[6] = {1, 4, 5, 2, 1, 2};
// myHappyNumbers[5] contains the number 2
// myHappyNumbers[6] returns an unexpected value
```

3.14 Program Control Structures

For controlling program flow, Arduino, just like C, has various structures available. `if-else` and `for` are the two most important ones.

`if`

checks whether a certain condition is met:

```
if (a > 10) {
// is executed when a is greater than 10
}
```

if-else

allows further control of program flow. Several conditions may be checked:

```
if (sensorValue < 500) {
  // this is executed when sensorValue is less than 500
}
else {
  // this is executed in all other cases
}
```

for

allows one to repeat an instruction or a block of instructions. The number of repetitions is not dependent on any other conditions.

```
for (initialization; condition; increment) {
    // statements
}
```

First, a starting condition, or *initialization*, is set. Each time the loop is executed, the *condition* is checked and the *increment* action is carried out. When *condition* is no longer met, the loop exits.

A loop that begins with

```
for (int i = 1; i <= 10; i++)
```

will run `i` through all of the values from 1 to 10, inclusive.

switch-case

Like the `if` statement, the `switch-case` structure controls program flow. With `switch-case`, however, several different cases can be tested within the braces.

A program will check each individual case and execute the branch where the test condition is TRUE.

`switch-case` has the following parameters:

var	variable to test
`default:`	This branch is optional, and executes only when no other conditions are matched.
`break;`	Without a `break`, the `switch` instruction would continue to search for the next matching condition, usually not desirable.

Example:

```
switch (value) {
  case 1:
  // this is executed when value == 1
  break;

  case 2:
  // this is executed when value == 2
  break;

  default:
  // this is executed in all other cases
}
```

while

`while` loops continue to execute as long as a loop condition is met. It is most useful when one does not know if, or how many times, the condition will be met.

```
while(condition) {
  // instructions
}
```

Example:

```
int dividend = 100;
int divisor = 6;
int quotient;

// calculate the result of a division (quotient)
// by repeatedly subtracting the divisor
while (dividend >= divisor) {
    dividend = dividend - divisor;
    quotient = quotient + 1;
}
```

do-while

The `do-while` loop is similar to the `while` loop, except that the condition is checked at the end of the loop. Therefore, the `do-while` loop will always execute at least once.

```
do {
  // instructions
}
while(condition);
```

Example:

```
// keep checking an input pin until it goes LOW;
do {
  val = digitalRead(inputPin);
}
while (val != LOW);
```

break

Using `break`, any `do-while`, `for` or `while` loop may be exited.

Example:

```
// keep checking an input pin until it goes LOW, unless another
// input pin goes HIGH, in which case, break out of the loop
do {
  val1 = digitalRead(inputPin1);
  val2 = digitalRead(inputPin2);

  if (val2 == HIGH) {
    break;
  }
} while (val1 != LOW);
```

3.15 Controlling Timing

Time delays were introduced earlier, in the *Blink* program. The following instructions are available for timing purposes.

`unsigned long millis()`

> Returns the number of milliseconds that have elapsed since the program began. This number goes as high as 2^{32} ms = 4,294,967,296 ms = 4,294,967.3 s = 1193.04 h = 49.71 days, so, after almost 50 days, it clocks over to 0 again.

`delay(ms)`

> causes a delay in program execution of *ms* milliseconds. The *Blink* sketch contains an example.

`delayMicroseconds(us)`

> causes a delay in program execution of *us* microseconds.

3 DEVELOPMENT ENVIRONMENT AND PROGRAMMING BASICS

3.16 Mathematical Functions

The most important mathematical functions have their usual meanings in Arduino:

`min(x,y); max(x,y); abs(x);`

return the minimum and maximum, respectively, of x and y, and the absolute value of x.

`sin(x); cos(x); tan(x);`

return the sine, cosine and tangent of x (expressed in radians), respectively.

`sqrt(x); pow(x, n); log(x);`

`sqrt(x)` returns the square root of x.

`pow(x, n)` raises x to the nth power, or x^n. Both x and n may be of the float type, so that common roots may be calculated using this function.

As in most programming languages, `log(x)` works somewhat differently to the usual mathematical convention, in that it calculates the logarithm of x using a base of e = 2.71828183. This function is often used to work with sensors that have an exponential response curve.

3.17 Random numbers

Random numbers are often needed, especially, for example, in the gaming field. The random numbers generated by the following functions are so-called pseudorandom numbers. These are not strictly random in the truest mathematical sense, as they are generated using a deterministic algorithm. For most applications in the gaming and hobbyist fields, however, their randomness is perfectly adequate.

`randomSeed(seed)`

starts the pseudorandom number generator. With this seed, the starting point in a pseudorandom sequence is set.

`long random(min, max)`

generates a pseudorandom number between min and max.

3.18 Extending the Options: Adding Libraries

Additional software libraries extend the functionality of the Arduino significantly. These very useful program packages are contained in Arduino's `libraries` subdirectory. After one includes the library using the IDE, additional functions become available. The following libraries are used in this book:

- CapSense – evaluation of capacitive measurement methods
- Dcf77 – analysis of time signals of the DCF77 standard
- EEPROM – use of external EEPROM devices
- IRremote – working with common infrared remote control signals
- LiquidCrystal – driving a liquid crystal display
- LiquidCrystal_I2C – driving a liquid crystal display using the I^{2}C bus
- PCF8583 – using the PCF8583 real-time clock chip
- Servo – controlling model servos
- SoftwareSerial – implementing a software-based serial interface
- SPI – using the SPI bus
- TimerOne – collection of routines for configuring the 16 bit hardware timer called Timer1
- Wire – library for the I^{2}C-bus

A library consists basically of three files:

- `keywords.txt`
- `LibraryName.cpp`
- `LibraryName.h`

and usually a subdirectory called `examples` is included, in which example programs for using the library are contained. The `keywords.txt` file is not strictly required. It simply defines which keywords should be highlighted in a different color in the main sketch.

For example, to install the TimerOne library, the zip file, `TimerOne-vX.zip`, must be downloaded from the link at

> www.arduino.cc/playground/Code/Timer1

A new subdirectory, `TimerOne`, must be created in Arduino's `libraries` subdirectory, and the three files in the Zip archive

- `TimerOne.cpp`
- `TimerOne.h`
- `keywords.txt`

are extracted to there, after which the library is available to the IDE after the IDE is restarted.

Other libraries may be installed in a similar manner.

4
Electronic Components and Low-Cost 'Freeduinos'

From the standpoint of operational safety, a good soldered PCB is hard to beat. For this reason, almost all electronic circuits in industry are delivered as circuit board assemblies. The Arduino also comes on a high-quality industrially-manufactured PCB.

However, in the world of circuit development, such PCBs have a serious drawback: when their development is completed, they cannot be easily modified. Even small adjustments or additions require a complete remanufacture. For this reason, there are prototype boards for laboratory purposes. Here, the components are laid out on a standard grid consisting of leaded points or circuit tracks and perhaps connected using hook-up wire. More information can be found in the sections that follow.

4.1 Breadboards: Simple and Effective without Soldering

A breadboard takes things one step back. Here, circuits are built without even any soldering. Sockets containing metal springs inside the board provide electrically conductive connections between electronic devices.

Figure 4.1 shows such a breadboard. The built-in connections are indicated with black lines. Of course, these boards do not approach the connection reliability of soldered circuit boards. Nevertheless, carefully constructed breadboard circuits can operate reliably for years. Therefore, this construction technique is not limited to short-lived experimental rigs. Under normal environmental conditions, i.e. no great temperature variations or excessive humidity, devices designed on breadboard can work just as well as professional electronics. In demanding applications such as in mobile devices or for in-vehicle use, a breadboard circuit is unlikely to be suitable. Due to the vibrations that occur in such environments, one cannot expect too high a level of reliability there, and one must resort to prototyping boards, or, better yet, etched printed circuit boards.

4 ELECTRONIC COMPONENTS AND LOW-COST 'FREEDUINOS'

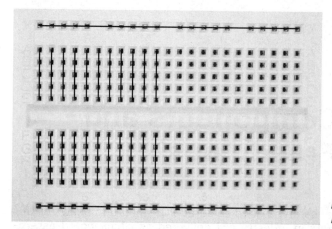

Figure 4.1:
Breadboard

4.2 Prototyping Boards: Durable Construction without Chemicals

If a particular circuit has been completely tested and is working properly, it may then be built on prototyping board. There are different versions available. The most common types are stripboard and perfboard. The former usually results in a circuit with fewer solder connections, although many circuit traces are likely to require severing. On the other hand, boards with individual solder pads require all electrical connections have to be individually created, so no manual severing is required on the board itself, but many more solder connections are required.

Figure 4.2:
Stripboard

Figure 4.3:
A Perfboard has Individual Pads

Even for simple circuits, one should always prepare a layout plan. On this plan, all of the necessary components are placed. The placement should be optimized so that the number and length of necessary interconnections is minimized.

After a final comparison with the circuit diagram, component placement may begin. It is advisable to always start with the small components such as capacitors and resistors. Larger components, such as power transistors and large electrolytic capacitors, follow. Finally, all of the necessary connections are made using tinned copper wire. A carefully constructed stripboard construction won't fare badly in comparison with a professional printed circuit board. If all solder joints are neatly done, stripboard manages to perform quite acceptably in terms of both durability and reliability.

Figure 4.4:
Stripboard Circuit

Another step in the direction of a professional PCB is the etched circuit board. This technique is certainly usable in amateur applications. However, the complexity is comparatively high. This technique is only justified when large quantities of identical PCBs are to be manufactured.

4.3 Low-Cost 'Freeduinos'

Naturally, one may also opt to construct the Arduino hardware oneself. The classic version of the controller with its DIL packaging and pin spacing of 1/10" is easily soldered onto stripboard. Since few other components are required aside from the crystal, a few capacitors and connectors, a suitable minimal Arduino board may be produced at less than a quarter the cost of the original version.

However, a programmer will still be required, that is, a USB-to-serial converter. This is difficult to make by hand, as the USB converter chip is only available in SMD versions. Since a suitable programmer costs only about half as much as an Arduino board, the savings would be negligible in any event if only one board is to be programmed. In a few special cases, DIY can still be interesting.

4.4 Arduino and Its Helpers: Basic Electronic Components

Aside from the Arduino, various other electronic components are required. At this stage, we won't go into a complete and detailed description of all of the components; there is a wealth of literature available [2], [3]. The essential characteristics, however, will be presented briefly. Of course, this description cannot be exhaustive. On occasion, it will still be necessary to seek further information, and the internet is obviously a good source thereof. Just about all of the common components have detailed datasheets available. These are usually easy to find using a search engine, and they may be downloaded free of charge.

4.4.1 USB Cable

The USB cable connects the Arduino to the PC. It serves not only for data transfer, but also supplies the microcontroller with power.

Note
Using an active hub with its own power supply, the PC's USB interface is largely protected against short circuits. In the case of a failure, only the relatively low-cost hub is affected, and the USB interface on the PC is not damaged.

4.4.2 Resistors

Resistors are non-polarized components, i.e. the direction of their installation is not important. Electronics professionals usually place resistors in a circuit so that the color code reads from left to right (or top to bottom).

Resistors with a 5% tolerance are coded using four color rings (see Table 3 for examples). Three of these indicate the resistance. The fourth is gold and, in our case, represents a tolerance of 5%. For metal film resistors with a 1% tolerance, five color bands are necessary, where the brown fifth band indicates the 1% tolerance. The tolerance indicator is made easier to recognize by it being somewhat thicker than the others.

For an introduction to microcontroller techniques, we get by with relatively few values. For LEDs or LED-based displays, 220 Ω resistors are often used. Values of 1 kΩ and 10 kΩ are also often needed. It is advisable to have a fair quantity of these (approx. 10 to 100 pieces per value). Even 100 kΩ resistors are useful, so it's a good idea to have a few of these available. If necessary, other values can be created by connecting resistors in parallel or in series (e.g. 500 Ω by connecting two 1 kΩ resistors in parallel, or 20 kΩ by connecting two 10 kΩ resistors in series). The following table shows the color combinations for our most-often-used resistors. A variety package is often a more cost-effective option than purchasing resistors individually.

Value	Color band combination for 5% tolerance	Color band combination for 1% tolerance
220 Ω	red-red-brown	red-red-black-black
1 kΩ	brown-black-red	brown-black-black-brown
10 kΩ	brown-black-orange	brown-black-black-red
100 kΩ	brown-black-yellow	brown-black-black-orange

Table 3: *Frequently-Used Resistor Values*

4.4.3 Capacitors

Much like resistors, capacitors are fairly robust devices. Simple capacitors are not polarized, but, with electrolytic types, care should be exercised. It is essential to observe the correct polarity with these, as reversing the polarity may destroy the component. Electrolytic capacitors thus always have an appropriate label to indicate polarity. To begin with, a few values suffice, and these are summarized in the following table.

Quantity	Value	Label
20	22 picofarad (pF)	22 or 22p
10	10 nanofarad (nF)	103 or 10n
10	100 nF	104 or 100n
10	1 microfarad (µF), 16 V	1 µF 16 V
10	10 µF, 16 V	10 µF 16 V
5	100 µF, 16 V	100 µF 16 V

Table 4: *Important Capacitor Values*

Just as with resistors, variety packs of capacitors are available. These usually work out cheaper than buying the components individually. With capacitors, it is usually not as critical that the exact values be used as it is with resistors. They are often just there to provide noise suppression as blocking capacitors. For this, the exact value is not critical. Instead of a 10 µF type, for example, a 33 µF may usually be substituted if that's what's available.

4.4.4 Potentiometers

Potentiometers (or 'pots' for short) are variable resistors. They usually consist of a resistive track with two connections, from which varying resistance values are achieved using a sliding contact. There are many types of pots. With some versions, wires are soldered directly to the terminals, while others may be directly insertable into a breadboard.

Tip
When connecting pots directly to the breadboard, it's advisable to connect the terminals in parallel to the plug springs. For one, this protects the springs, and the pot can also be inserted with far less effort. A pair of pliers may be used to twist the terminals around 90 degrees in order to achieve this. Typical resistance values for potentiometers are 10 kΩ and 100 kΩ.

4.4.5 LEDs

Light-emitting diodes (LEDs) are among the most important components for the projects in this book. Besides the classic 5 mm package types, 3 mm variants are a great option, as these can be placed right next to each other on a breadboard so that the available space is optimized. LEDs with a rated current of 20 mA are still widely used, but the much more efficient 2 mA types are readily available. These should always be preferred, as they place far less load on the port pins.

LEDs light up only when they are installed at the correct polarity. If an LED does not light up as expected, double-check the polarity. The cathode of the LED is indicated by a flattening on the plastic housing, and it should always be connected to the negative voltage potential. The anode is usually the longest of the two terminals.

Another tip
So-called **5 V LEDs** already contain built-in resistors. For this reason, the external resistors previously required are made redundant. This makes the design of circuits on small breadboards much simpler. Another advantage to this type of LED is that it is practically invincible when used with an Arduino, as no voltages higher than 5 V are encountered. Thus, these special LEDs, in contrast to the conventional types, cannot be overloaded.

REMEMBER
When connecting standard LEDs to a standard 5 V source, a resistor is always required.

4.4.6 RGB LEDs

Multicolor LEDs are a recent development in the field of optoelectronics. With them, it's possible to produce just about any color in the visible spectrum. This is made possible through the use of three individually-colored LEDs, red, green and blue, adjacent to each other in a single package. These LEDs are often referred to as RGB LEDs. They are becoming more common as high-intensity large display components, used in giant arrays, for advertising purposes or for information boards.

The cathode is usually the longest terminal, and, to the left and right, are the anodes for the red, green and blue LED elements.

Note
 Each individual element in an RGB LED also requires that there be a resistor in series for over-current protection. This means that you will require three resistors per RGB LED.

4.4.7 Switches

Switches are available in various designs. There are also types that may be inserted directly into breadboard connections. Keys are pressed mainly to affect program flow. Usually, the key sinks the port to ground, i.e. when pressing the key, the microcontroller input is connected to the ground potential. When the switch is inactive, the input signal would normally be unstable and undefined, but a pull-up resistor (typically between 1 kΩ and 10 kΩ) to 5 V is used to prevent this. When the switch is open, the input port is held at its high potential. However, the Arduino has internal pull-up resistors that may be enabled.

4.4.8 Silicon Diodes

Diodes behave like one-way streets, allowing current to pass in only one direction. Standard silicon diodes, such as the 1N4148 types, should be driven with a maximum of 50 mA, i.e., just as with LEDs, these should never be connected directly to a 5 V power supply. The cathode of this type of diode is marked with a black ring. The diode conducts electricity as long as the anode is connected to positive potential and the cathode to negative potential.

4.4.9 Transistors

A normal processor pin can drive a maximum of a few tens of milliamps. There are other limitations to consider along with that. With many types of controllers, the sum of the output current that the controller is able to deliver may not exceed a certain value (e.g. 200 mA). When using several pins simultaneously, this value can easily be exceeded. Once larger currents have to be switched, the use of transistors should be considered. The most important characteristic of transistors is that they amplify currents. With the small base currents supplied by the Arduino, large collector currents are controllable via the transistor. For this, the collector (C) must be connected to the positive potential, and the emitter (E) to the negative. As soon as a sufficient base current flows, the transistor begins to conduct. Both the collector and the base must also be protected from overload by using suitable resistors.

4 ELECTRONIC COMPONENTS AND LOW-COST 'FREEDUINOS'

5 Hello World

One of the easiest microcontroller applications is switching LEDs on and off. This is a task especially suited to microcontrollers. Classic LEDs require a current of around 20 mA. These are easily driven by a microcontroller pin. Newer low-current types exist that require only 2 mA. In this first major project, ten LEDs are controlled. Figure 5.1 shows the construction of the circuit.

The 5 V LEDs mentioned in section 4.4.5 are ideal for this. These may be connected directly to the Arduino's outputs. Then, the current limiting resistors shown in Figure 5.1 are not required, as these are already integrated into the LEDs.

5.1 Cut to the Chaser

If several LEDs are used, running light 'chase' effects are easily achieved, e.g. for case modding or on railway models.

In the program, the LED pin numbers are initially declared as an array. The `direction` variable defines the direction in which the light appears to move, while the `LED` variable contains the port number of the currently activated LED. During setup, all wired ports are initially defined as outputs and set to low potential.

In the main loop, the first LED is lit using

```
digitalWrite(ledPin[LED], HIGH);
```

After a 50 ms delay, the same LED is turned off again, and then the `LED` variable is increased by the value of `direction`. So, should `direction == 1`, the point of light would move in the direction of the higher port numbers, i.e. to the left. If `direction == -1`, it moves in the other direction.

Using two `if` statements, the end of the light chain is detected at LED0 or LED9, and the direction is inverted.

5 HELLO WORLD

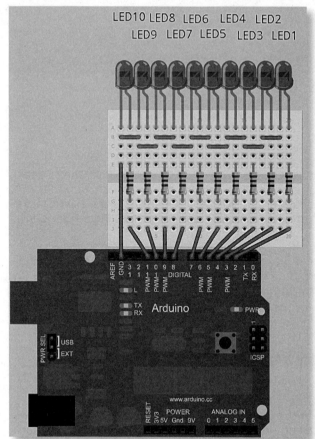

Figure 5.1:
LED Chaser

Listing 5-1: LED Chaser

```
// Listing 5-1
// LED Chaser

const byte ledPin[] = {
  2, 3, 4, 5, 6, 7, 8, 9, 10, 11
};

int drctn = 1;
int led = 0;

void setup() {
  // all LED pins as outputs and LEDs off
  for (int x = 0; x < 10; x++) {
    pinMode(ledPin[x], OUTPUT);
    digitalWrite(ledPin[x], LOW);
  }
}
```

```
void loop() {
  // turn on current LED
   digitalWrite(ledPin[led], HIGH);
   delay(50);

  // turn off current LED
  digitalWrite(ledPin[led], LOW);
  led += drctn;

  // change direction if we reach the end
  if (9 == led) {
    drctn = -1;
  }

  // or the beginning
  if (0 == led) {
    drctn = 1;
  }
}
```

Extensions and exercises

- Instead of a single point of light, let two LEDs run directly next to each other.

- Try and make one light point move from right to left while another moves from left to right, simultaneously.

- Make an animated demo: first, let a light point run from left to right, then blink all of the LEDs in one-second intervals, then run a light from right to left, then let five LEDs light simultaneously while the other five remain dark. Finally, repeat this all in an infinite loop.

5.2 It Gets Brighter: Controlling Power LEDs

As mentioned before, a single Arduino pin should not be used to drive more than 20 mA. This is fine for a single standard LED, but modern ultra-bright power LEDs often require in excess of 300 mA to do their job. Here, a transistor can help. The following diagram shows a circuit capable of running power LEDs of up to 350 mA.

5 HELLO WORLD

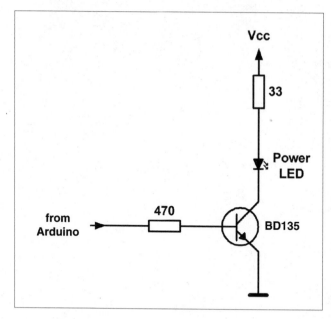

Figure 5.2: Controlling a Power LED

An example transistor suitable for this would be a BD135, as it is able to handle currents of over 1 A.

5.3 POVino: Persistence-of-Vision Display

A very interesting way to display short text messages is demonstrated in this project. Here, we take advantage of a special characteristic of the human visual system, which is that short impulses of light exhibit an 'afterglow'. This phenomenon is easily demonstrated using a flashlight. When the flashlight is moved around quickly in a dark environment, it seems as if the point of light leaves a trail behind it. In this way, one may draw simple geometric figures, such as circles or rectangles, in the dark.

Now, if we turn individual LEDs on and off within a row of LEDs very rapidly, while simultaneously shaking the circuit about in a dark room, numbers and letters may be generated. This is known as a persistence-of-vision, or POV, display. By equipping the Arduino with a row of LEDs, a simple device for emergencies may be assembled. Using the following sketch, one may draw the word 'HELP' in the air.

The corresponding circuit is shown in Figure 5.3.

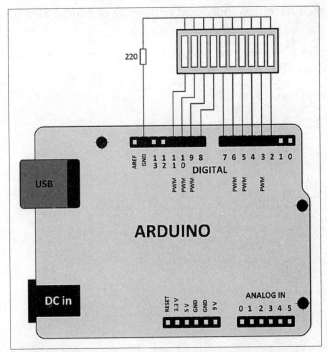

Figure 5.3:
Circuit Diagram for POV display

Figure 5.4:
Construction of the POV Display

Figure 5.5:
POV 'HELP' Display

5 HELLO WORLD

Instead of the LED bar demonstrated here, ten individual LEDs may of course also be used.

In the program, the individual letters are stored as arrays. All of the letters are represented as part of a virtual 5x3 matrix. The letter 'E' looks like this:

```
111   • • •
100   •
111   • • •
100   •
111   • • •
```

So, in the individual columns:

```
11111
10101
10101
```

and thus:

```
int E[] = {1,1,1,1,1, 1,0,1,0,1, 1,0,1,0,1};
```

The duration of the character's illumination and the duration of the dark periods between characters is specified in the sketch.

In the `writeChar()` function, the letter matrices are sent to the LED array in succession. By shaking the assembly rapidly, the word appears, as shown in Figure 5.5.

Listing 5-2: *Persistence of Vision*

```
// Listing 5-2
// Persistence of Vision

const int H[] = {1,1,1,1,1, 0,0,1,0,0, 1,1,1,1,1};
const int E[] = {1,1,1,1,1, 1,0,1,0,1, 1,0,1,0,1};
const int L[] = {1,1,1,1,1, 0,0,0,0,1, 0,0,0,0,1};
const int P[] = {1,1,1,1,1, 1,0,1,0,0, 1,1,1,0,0};

const byte darkTime = 5, onTime = 3;
```

```
  const byte startPort = 2, charLength = 3, charHeight = 5;

  void setup() {
    for (int i = startPort; i < startPort + charHeight; i++) {
      pinMode(i, OUTPUT);
    }
  }

  void writeChar(const int character[]) {
    for (int x = 0; x <= (2 * charHeight); x += charHeight) {
      for (int y = 0; y < charHeight; y++) {
        digitalWrite(y + startPort, character[y + x]);
      }
      delay(onTime);
    }

    for (int y = 0; y < charHeight; y++) {
      digitalWrite(y + startPort, 0);
    }
    delay(darkTime);
  }

  void loop() {
    writeChar(H);
    writeChar(E);
    writeChar(L);
    writeChar(P);
  }
```

The display is the most important interface between humans and electronics. In recent years, there has been tremendous progress in this area. At the end of the 90s, CRT monitors, with their flickering images, low resolutions and high energy consumption, dominated the computer field. With the advent of flat screens, the situation improved dramatically. There are now full HD monitors available in diagonal sizes of over a meter.

Mobile devices have also revolutionized display technology. Without brilliant, ultra-flat and robust displays, devices such as the latest iPods, iPads, digital cameras and smart phones would be unthinkable.

In principle, even high-resolution color displays are driven by microcontrollers, but this would be a little overwhelming for the Arduino Uno. Still, various attractive displays are possible with smaller controllers. From simple bar graphs to 7-segment displays, LED dot matrix displays to liquid crystal text displays, the technology leaves little to be desired. Obviously applications such as the display and visualization of instrumentation and other data are more relevant to microcontrollers, while high-resolution graphics and speedy video games remain the domain of stronger processors.

5 HELLO WORLD

6
Displays and Display Techniques

6.1 Bar Graph Display: The Classic for Measurement Applications

A simple display type was used earlier in our POV project: the LED bar graph. This consists of individual rectangular LEDs packaged together to form a single row. Besides POV experiments, these are capable of other optical effects, such as bands of light or chase effects.

When using LED bar graph displays, one should always ensure that appropriate resistors are used. If only a single LED is active at a time, then a single, common resistor for all LED elements is sufficient. If several LEDs are illuminated simultaneously, there is the disadvantage that the LEDs get dimmer as more of them are illuminated at once. This happens because, at increased current, the voltage drop over the common resistor increases as well, which leaves less voltage for the LEDs to work with.

For a POV application, this isn't a real problem. For bar graph displays, however, one should always use an individual resistor per LED element. In this way, all of the LEDs will be equally bright, regardless of how many are illuminated.

Further details and examples of the use of bar graph displays may be found in Chapter 7.

Extensions and exercises
Adapt the running light program from Section 5.1 to the bar graph display used in the POV project.

6.2 Simple and Cheap: 7-Segment Displays

For displaying numbers, 7-segment displays are best suited. They consist of seven (or eight, counting the decimal point) LED elements. These segments are arranged in the form of the number '8'. By activating specific segments, the digits from '0' to '9' may be represented with good legibility. In addition, a few capital letters may be displayed (A, C, E, F, etc.). In other cases, such as with the capital letter 'B', the display would be indistinguishable from an 8. Here, one may resort to lower-case letters. Other letters, such as 'M' and 'V', are difficult to represent unambiguously. For this, 16-segment displays are available.

However, these are relatively uncommon and therefore expensive, so they won't be described any further here. Another option for displaying alphabetical characters, the dot matrix display, is covered in a later section.

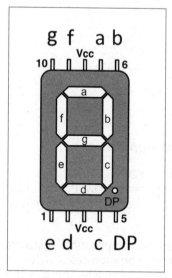

Figure 6.1:
Typical Pin Layout on a Single 7-Segment Display

The wiring of a common-cathode 7-segment display is shown in Figure 6.2. Since a numeric digit requires several segments to be lit simultaneously, one might expect an individual (approx. 220 Ω) resistor for each segment, as per the stipulation in Section 6.1. However, as will be explained in more detail in the description of the control software, by using so-called multiplexing, a single resistor suffices.

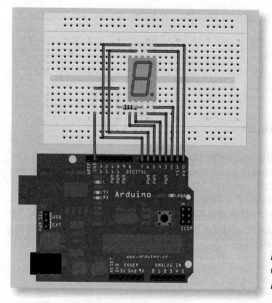

Figure 6.2:
Control of a Single 7-Segment Display.

SIMPLE AND CHEAP: 7-SEGMENT DISPLAYS 6.2

Arduino pin	0	1	2	3	4	5	6	7
Display pin	6 (b)	7 (a)	5 (DP)	4 (c)	2 (d)	1 (e)	10 (f)	9 (g)

Table 5:
Arduino-to-7-Segment-Display Pin Assignments

Figure 6.3:
7-Segment Display in Action

Listing 6-1: *Multiplexed 7-Segment Display*

```
// Listing 6-1

// Multiplexed 7-Segment Display

// 7-segment codes
const int numbers[10][8] = {
  // b, a, DP, c, d, e, f, g
  {1, 1, 0, 1, 1, 1, 1, 0}, // 0
  {1, 0, 0, 1, 0, 0, 0, 0}, // 1
  {1, 1, 0, 0, 1, 1, 0, 1}, // 2
  {1, 1, 0, 1, 1, 0, 0, 1}, // 3
  {1, 0, 0, 1, 0, 0, 1, 1}, // 4
  {0, 1, 0, 1, 1, 0, 1, 1}, // 5
  {0, 1, 0, 1, 1, 1, 1, 1}, // 6
  {1, 1, 0, 1, 0, 0, 0, 0}, // 7
  {1, 1, 0, 1, 1, 1, 1, 1}, // 8
  {1, 1, 0, 1, 1, 0, 1, 1}  // 9
};

void setup() {
  for (int i = 0; i <= 7; i++) {
    pinMode(i, OUTPUT);
  }
}
```

```
void loop() {
  for (int n = 0; n <= 9; n++) {
    for (int i = 0; i <= 30; i++) {
      updateDisplay(n);
    }
  }
}
f
void updateDisplay(int digit) {
  for (int i = 0; i <= 7; i++) {
    digitalWrite(i, LOW); // clear all segments
  }

  // activate one segment after another
  for (int i = 0; i <= 7; i++) {
    if (1 == numbers[digit][i]) {
      digitalWrite(i, HIGH); // set segment
      delay(3);
      digitalWrite(i, LOW); // clear segment
    }
    else {
      delay(3); // time equalization
    }
  }
}
```

Should the pin assignment of the 7-segment display not match the one used here, one may adapt either the circuit or simply the pin table array contained in the software.

Although it's possible to drive all segments simultaneously on a single-digit 7-segment display, the multiplex method was used here. This means that, at any given moment, only a single segment is active. This also relies on persistence of vision, as the rapid repeated activation of a succession of segments creates the impression of a single, consistent digit.

6.3 4-Digit, 7-Segment Displays: A Key Component for Instrumentation

Using a single 7-segment display, it is possible to indicate simple operating conditions, such as a washing machine's cycle status. Alternately, a series of digits may be displayed sequentially. For most applications, however, more digits are needed.

A 4-digit display is very versatile. Should a colon also be included, then we have a very versatile display at hand. If both dots in the colon are activated, this is suited to the display of the time, or as minute and second displays for timers and countdowns. If only the lower dot is used, then one may represent values to two decimal places. In this way, many practical devices, such as voltmeters and ammeters, ohmmeters and even thermometers are possible.

Figure 6.5 shows how a common-anode 4-digit, 7-segment display may be connected

4-DIGIT, 7-SEGMENT DISPLAYS: A KEY COMPONENT FOR INSTRUMENTATION 6.3

to the Arduino. Since twelve interconnections and four resistors are required, the usual construction diagram is not used; rather, the circuit is represented using a classic schematic diagram.

Arduino Pin	0	1	2	3	4	5	6	7	A0	A1	A2	A3
4x7-segment Pin	1 (e)	2 (d)	3 (DP)	4 (c)	5 (g)	7 (b)	10 (f)	11 (a)	12 (D1)	9 (D2)	8 (D3)	6 (D4)

Table 6: *Arduino-to-4x7-segment display pin assignments*

In addition to the segments (a – g), the common anode for Digits 1 – 4 (left to right) are indicated as D1, D2, D3 and D4, respectively.

A nested multiplex method is used here, in which only a single segment of a single one of the four digits is active at a time. This reduces the component count to a minimum.

It is common to see multi-digit LED displays making use of the regular multiplex method, i.e., all relevant segments of each individual digit will be active simultaneously. However, due to the higher current consumption in that configuration, four additional transistors and seven resistors would also have been required.

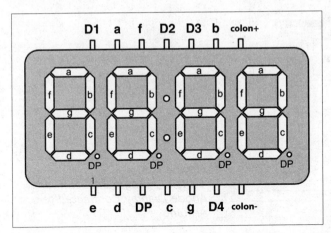

Figure 6.4:
Integrated 4-Digit, 7-Segment Numeric Display

The software for controlling the display is divided into two parts. The main program, *Test 4x7 Display* (Listing 6-2), is kept short. Two libraries are included, an initialization routine is called and the number sequence '1234' is sent to the display using the numberOutput() function.

The second part of the control software, *LED_display.h* (Listing 6-3), is a driver. It is the driver that determines exactly how a digit is represented on the display. With the two functions defined in this routine,

```
void initLedDdisplay()
```

6 DISPLAYS AND DISPLAY TECHNIQUES

and

```
void numberOutput(uint16_t number)
```

control of the display is very easy. The first function initializes the 4x7-segment display. The second displays a four-digit number, i.e. a value between 0000 and 9999, inclusive.

In this sketch, the TimerOne library is used, so it must be installed.

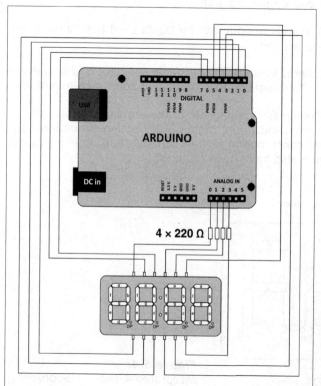

Figure 6.5:
4-Digit, 7-Segment Common Anode Display Connected to Arduino

Listing 6-2: Test 4x7 LED Display

```
// Listing 6-2
// Test 4x7 LED Display
#include <TimerOne.h>
#include "LedDisplay.h"   // include our display driver

void setup() {
  initLedDisplay();
}

void loop() {
```

```c
    numberOutput(1234);
}
```

Listing 6-3: *LedDisplay.h*

```c
// Listing 6-3
// LedDisplay.h
#include <avr/io.h>
#include <avr/interrupt.h>
#include <util/delay.h>

// 7-segment codes
const int numbers[10][8] = {
  // e, d, DP, c, g, b, f, a
  {0, 0, 1, 0, 1, 0, 0, 0}, // 0
  {1, 1, 1, 0, 1, 0, 1, 1}, // 1
  {0, 0, 1, 1, 0, 0, 1, 0}, // 2
  {1, 0, 1, 0, 0, 0, 1, 0}, // 3
  {1, 1, 1, 0, 0, 0, 0, 1}, // 4
  {1, 0, 1, 0, 0, 1, 0, 0}, // 5
  {0, 0, 1, 0, 0, 1, 0, 0}, // 6
  {1, 1, 1, 0, 1, 0, 1, 0}, // 7
  {0, 0, 1, 0, 0, 0, 0, 0}, // 8
  {1, 0, 1, 0, 0, 0, 0, 0}  // 9
};

// store single digits
volatile int D1, D2, D3, D4;

volatile uint8_t activeDigit = 0;

#define sbi(PORT, bit) (PORT |= (1 << bit)) // set bit in PORT
#define cbi(PORT, bit) (PORT &= ~(1 << bit)) // clear bit in PORT

// Write number at indicated position to display
void digit(int value, int pin) {
  sbi(PORTC, pin); // pin high => current digit on
  // cbi(PORTC, pin); // use for common cathode
  PORTD = 0b11111111;
  // PORTD = 0b00000000; // use for common cathode

    // activate one segment after another
    for (int i = 0; i <= 7; i++) {
      if (0 == numbers[value][i]) {
        cbi(PORTD, (i)); // set segment
        // sbi(PORTD, (i)); // use for common cathode
        _delay_ms(1);
        sbi(PORTD, (i)); // clear segment
        // cbi(PORTD, (i)); // use for common cathode
```

```c
      }
    }
  }

  // distribute number to the digits
  void numberOutput (uint16_t number) {
    D1 = number/1000; number %= 1000;
    D2 = number/100;  number %= 100;
    D3 = number/10;   number %= 10;
    D4 = number;
  }

  void updateDisplay() {
    PORTC = 0b00000000; // all digits off
    // PORTC = 0b00001111; // use for common cathode

    if (0 == activeDigit) {
      digit(D1, PORTC0);
    }

    if (1 == activeDigit) {
      digit(D2, PORTC1);
    }

    if (2 == activeDigit) {
      digit(D3, PORTC2);
    }

    if (3 == activeDigit) {
      digit(D4, PORTC3);
    }

    activeDigit++;
    if (4 == activeDigit) {
      activeDigit = 0;
    }
  }

  void initLedDisplay() {
    DDRD = 0b11111111;
    DDRC = 0b00001111;
    // Interrupt every 1000 us = 1 ms
    Timer1.initialize(1000);
    Timer1.attachInterrupt(updateDisplay);
  }
```

6.4 Mini Monitor for Signs and Graphics: The LED Dot Matrix

Along with 7-segment displays, dot matrix LED displays have gained widespread acceptance. The advantage of these is their significantly higher flexibility. On 7-segment displays, only numbers and a limited number of letters may be unambiguously displayed. Dot matrix displays, on the other hand, allow the display of all alphanumerical characters and even simple graphics and icons. Dot matrix displays are particularly popular in Asia, due to their ability to represent East Asian characters.

Figure 6.6 shows the wiring of a dot matrix circuit. By approaching the relatively complex wiring systematically, one shouldn't encounter any problems. Table 7 provides further assistance. Begin by connecting Arduino Pin 1 to Pin 6 of the display, then Arduino Pin 2 to display Pin 5, and so forth. Don't forget the resistors required at Arduino pins 8 – 12!

Arduino	1	2	3	4	5	6	7	8	9	10	11	12
Dot matrix	6	5	4	9	2	11	12	8 (via 220 Ω)	7 (via 220 Ω)	10 (via 220 Ω)	3 (via 220 Ω)	1 (via 220 Ω)

Table 7: *Pin Assignment for Connecting Arduino to Dot-Matrix Display*

The following program brings up some graphical characters on the display.

Listing 6-4: *Dot Matrix Graphic Display*

```
// Listing 6-4
// Dot Matrix Graphic Display

// Row Col 1 2 3 4 5 6 7
// 1          o o o o o o o
// 2          o o o o o o o
// 3          o o o o o o o
// 4          o o o o o o o
// 5          o o o o o o o

const int maxDelay = 500;

// define column and row pins
const int col[7] = {7, 6, 5, 4, 3, 2, 1};
const int row[5] = {12, 11, 10, 9, 8};

// define graphics numbers as 3x5 sub-matrices
const int smiley[7][5] = {
    {0, 0, 0, 0, 0},
    {0, 1, 0, 1, 0},
    {0, 0, 0, 0, 0},
    {0, 0, 1, 0, 0},
    {1, 0, 0, 0, 1},
```

```
    {0, 1, 1, 1, 0},
    {0, 0, 0, 0, 0}
};

const int heart[7][5] = {
    {0, 0, 0, 0, 0},
    {0, 1, 0, 1, 0},
    {1, 0, 1, 0, 1},
    {1, 0, 0, 0, 1},
    {0, 1, 0, 1, 0},
    {0, 0, 1, 0, 0},
    {0, 0, 0, 0, 0}
};

const int diamond[7][5] = {
    {0, 0, 0, 0, 0},
    {0, 0, 1, 0, 0},
    {0, 1, 0, 1, 0},
    {1, 0, 0, 0, 1},
    {0, 1, 0, 1, 0},
    {0, 0, 1, 0, 0},
    {0, 0, 0, 0, 0}
};

void setup() {
  for (int i = 0; i < 7; i++) {
    pinMode(col[i], OUTPUT);
  }
  for (int i = 0; i < 5; i++) {
    pinMode(row[i], OUTPUT);
  }
}

void loop() {
  // show smiley
  for (int del = 0; del < maxDelay; del++) {
    for (int x = 0; x < 5; x++) {
      for (int y = 0; y < 7; y++) {
        digitalWrite(col[y], smiley[y][x]);
      }
      // set current column to 0
      digitalWrite(row[x], 0);
      delay(1);
      // reset column
      digitalWrite(row[x], 1);
    }
  }

  // show heart
  for (int del = 0; del < maxDelay; del++) {
    for (int x = 0; x < 5; x++) {
```

```
      for (int y = 0; y < 7; y++) {
        digitalWrite(col[y], heart[y][x]);
      }
      // set current column to 0
      digitalWrite(row[x], 0);
      delay(1);
      // reset column
      digitalWrite(row[x], 1);
    }
  }

  // show diamond
  for (int del = 0; del < maxDelay; del++) {
    for (int x = 0; x < 5; x++) {
      for (int y = 0; y < 7; y++) {
        digitalWrite(col[y], heart[y][x]);
      }
      // set current column to 0
      digitalWrite(row[x], 0);
      delay(1);
      // reset column
      digitalWrite(row[x], 1);
    }
  }
}
```

6 DISPLAYS AND DISPLAY TECHNIQUES

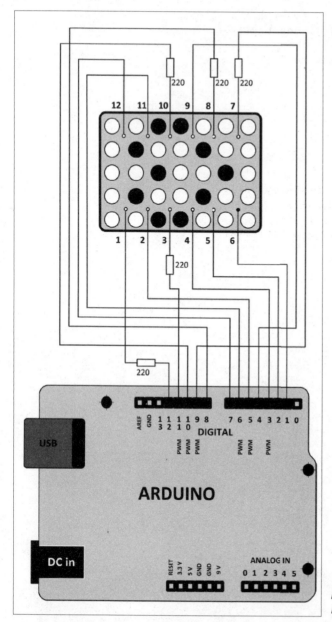

Figure 6.6:
Mini Graphical Display Circuit

6.5 Dot Matrix Display as a Two-Digit Digital Display

When used in 'landscape' orientation, a 5x7 matrix becomes a 7x5, which allows the representation of two digits. In this way, the display is well suited for two-digit displays of voltage, counter status, etc. The assembly is shown in Figure 6.7.

DOT MATRIX DISPLAY AS A TWO-DIGIT DIGITAL DISPLAY 6.5

Figure 6.7:
Dot Matrix Display as a Two-Digit Digital Display

Listing 6-5: *Dot Matrix Counter Display*

```
// Listing 6-5
// Dot Matrix Counter Display

// Row   Col 1 2 3 4 5 6 7
//  1        o o o o o o o
//  2        o o o o o o o
//  3        o o o o o o o
//  4        o o o o o o o
//  5        o o o o o o o

int number; // number on display

// define columnn and row pins
const int col[7] = {7, 6, 5, 4, 3, 2, 1};
const int row[5] = {12, 11, 10, 9, 8};
const int n[][3] = {
  {31, 17, 31}, // 0
  { 0, 31,  2}, // 1
  {23, 21, 29}, // 2
  {31, 21, 21}, // 3
  {31,  4,  7}, // 4
  {29, 21, 23}, // 5
  {29, 21, 31}, // 6
  {31,  1,  1}, // 7
  {31, 21, 31}, // 8
  {31, 21, 23}  // 9
};

void setup() {
  for (int i = 0; i < 7; i++) {
    pinMode(col[i], OUTPUT);
  }
```

```
    for (int i = 0; i < 5; i++) {
      pinMode(row[i], OUTPUT);
    }
  }

  void loop() {
    // count from 00 to 99
    for (number = 0; number <100; number++) {
      for(int n = 1; n < 50; n++) {
        showMatrix(number);
      }
    }
  }

  void showMatrix(int displayNumber) {
    // ones digit
    for (int x = 1; x <= 3; x++) {
      digitalWrite(x, HIGH);
      PORTB = ~n[displayNumber % 10][x - 1];
      delay(1);
      digitalWrite(x, LOW);
    }
    // tens digit
    for (int x = 5; x <= 7; x++) {
      digitalWrite(x, HIGH);
      PORTB = ~n[displayNumber / 10][x - 5];
      delay(1);
      digitalWrite(x, LOW);
    }
  }
```

6.6 Micro Learns to Write: Alphanumeric Display

Aside from the representation of graphical symbols and numbers, dot matrix displays are especially suited to the display of alphabetical characters. They are increasingly being used on buses, at bus stops and on trains. Large displays with hundreds of dots are usually seen there. If only a single, small matrix is available, one may find it helpful to display two characters alongside each other. In this way, longer text messages might also be displayed on larger displays.

6.6 MICRO LEARNS TO WRITE: ALPHANUMERIC DISPLAY

Figure 6.8:
Dot Matrix Display of an Alphabetical Character

In the program, the letters to be displayed are declared as two-dimensional arrays. Then, the arrays are sent to the display sequentially.

Listing 6-6: *Dot Matrix Character Display*

```
// Listing 6-6
// Dot Matrix Character Display

//       Col 1 2 3 4 5 6 7
// Row
// 1         o o o o o o
// 2         o o o o o o
// 3         o o o o o o
// 4         o o o o o o
// 5         o o o o o o

const int maxDelay =100;      // for "delays"

// define column and row
const int col[7] = {7, 6, 5, 4, 3, 2, 1};
const int row[5] = {12, 11, 10, 9, 8};

// define graphics characters as 5x7 matrices
int H[7][5] = {
```

```
    {1, 0, 0, 0, 1},
    {1, 0, 0, 0, 1},
    {1, 0, 0, 0, 1},
    {1, 1, 1, 1, 1},
    {1, 0, 0, 0, 1},
    {1, 0, 0, 0, 1},
    {1, 0, 0, 0, 1}
};

int E[7][5] = {
    {1, 1, 1, 1, 1},
    {1, 0, 0, 0, 0},
    {1, 0, 0, 0, 0},
    {1, 1, 1, 1, 0},
    {1, 0, 0, 0, 0},
    {1, 0, 0, 0, 0},
    {1, 1, 1, 1, 1}
};

int L[7][5] = {
    {1, 0, 0, 0, 0},
    {1, 0, 0, 0, 0},
    {1, 0, 0, 0, 0},
    {1, 0, 0, 0, 0},
    {1, 0, 0, 0, 0},
    {1, 0, 0, 0, 0},
    {1, 1, 1, 1, 1}
};

int O[7][5] = {
    {0, 1, 1, 1, 0},
    {1, 0, 0, 0, 1},
    {1, 0, 0, 0, 1},
    {1, 0, 0, 0, 1},
    {1, 0, 0, 0, 1},
    {1, 0, 0, 0, 1},
    {0, 1, 1, 1, 0}
};

void setup() {
  for (int i = 0; i < 7; i++) {
    pinMode(col[i], OUTPUT);
  }
  for (int i = 0; i < 5; i++) {
    pinMode(row[i], OUTPUT);
  }
}

void loop() {
  // H
  for (int del = 1; del < maxDelay; del++) {
```

```
      for (int x = 0; x < 5; x++) {
        for (int y = 0; y < 7; y++) {
          digitalWrite(col[y], H[y][x]);
        }
        // set current column to 0
        digitalWrite(row[x], 0);
        delay(1);
        // reset column
        digitalWrite(row[x], 1);
      }
    }
    delay(100);

    // E
    for (int del = 1; del < maxDelay; del++) {
      for (int x = 0; x < 5; x++) {
        for (int y = 0; y < 7; y++) {
          digitalWrite(col[y], E[y][x]);
        }
        // set current column to 0
        digitalWrite(row[x], 0);
        delay(1);
        // reset column
        digitalWrite(row[x], 1);
      }
    }
    delay(100);

    // display the letter 'L' twice
    for (int i = 0; i < 2; i++) {
      // L
      for (int del = 1; del < maxDelay; del++) {
        for (int x = 0; x < 5; x++) {
          for (int y = 0; y < 7; y++) {
            digitalWrite(col[y], L[y][x]);
          }
          // set current column to 0
          digitalWrite(row[x], 0);
          delay(1);
          // reset column
          digitalWrite(row[x], 1);
        }
      }
      delay(100);
    }

    // O
    for (int del = 1; del < maxDelay; del++) {
      for (int x = 0; x < 5; x++) {
        for (int y = 0; y < 7; y++) {
          digitalWrite(col[y], O[y][x]);
```

```
      }
      // set current column to 0
      digitalWrite(row[x], 0);
      delay(1);
      // reset column
      digitalWrite(row[x], 1);
    }
  }
  delay(500);
}
```

6.7 The LCD

The LCD is certainly among the most important and commonly used microcontroller peripherals. With it, letters, numbers and other symbols may be displayed. This makes it possible to realize very professional small devices for various applications. Temperature, voltage or other metrics may be displayed just as easily as time of day or other short messages.

In contrast to LED dot matrix displays, LCDs are almost exclusively available with integrated controllers. The reason for this is that a very large number of pixels must be driven. Even a small display consisting of two rows of 16 characters each would require controlling over 1,000 pixels. This would far exceed the typical capability of a typical AVR microcontroller. The HD44780 display controller has emerged as a quasi-standard. This may be operated over 16 wires, and the associated pin assignment is shown in the following table.

Pin	Symbol	Function
1	VSS	power supply 0 V (GND)
2	VDD	power supply +5 V
3	VEE	contrast adjustment (analog voltage)
4	RS	Register Select (instruction / data)
5	R/W	'HIGH' = read, 'LOW' = write
6	E	Enable (falling edge)
7	D0	Display data, LSB
8	D1	Display data
9	D2	Display data
10	D3	Display data
11	D4	Display data
12	D5	Display data
13	D6	Display data
14	D7	Display data, MSB
15	A	Backlight anode (+5 V)
16	K	Backlight cathode (GND)

Table 8: HD44780-Compatible LCD display – Pin Assignment

Contrast voltage VEE should ideally be connected between GND and 5 V via a 10 kΩ potentiometer, allowing the display contrast to be adjusted. Alternately, the pin may simply be tied to ground via a 1 kΩ resistor. With many types of displays, this simplified connection results in a decent contrast ratio.

The HD44780 has an 8-bit wide data bus (D0 – D7), but, to save precious processor pins, it is possible to configure the device for 4-bit bus operation, in which the 8-bit data is transferred sequentially via two 4-bit nibbles over D4 – D7. In this case, data lines D0 – D3 are not required and may be left unconnected. The only disadvantage of this method is the somewhat lower transmission speed. In most applications, this is of little consequence. For this reason, HD44780 modules are almost exclusively deployed in their 4-bit mode.

Thanks to the frequent use of LC displays with microcontrollers, the Arduino IDE comes with a suitable library, `LiquidCrystal.h`. The LCD's pin assignments need merely be specified using initialization member function

```
LiquidCrystal lcd(12, 11, 5, 4, 3, 2);
```

Then, using the

```
lcd.print("Test");
```

function, text is sent to the display.

There is one small drawback to the current version of the LCD library, however. Although the function

```
lcd.begin(16, 4);
```

allows the display format to be set to four rows of 16 characters, a problem arises with that configuration. Despite the parameters 16 and 4, a 20-character display is actually expected by the library. The last four characters on a 16-character line then appear as the first four characters on the next line. In sketches, however, a bit of appropriate programming gets around this easily enough.

Arduino	GND	5 V	GND (via 1 kΩ)	12	GND	11	5	4	3	2	5 V	GND
LCD	1 (VSS)	2 (VDD)	3 (VEE)	4 (RS)	5 (R/W)	6 (E)	11 (D4)	12 (D5)	13 (D6)	14 (D7)	15 (A)	16 (K)

Table 9: Arduino-to-LCD Connections

6 DISPLAYS AND DISPLAY TECHNIQUES

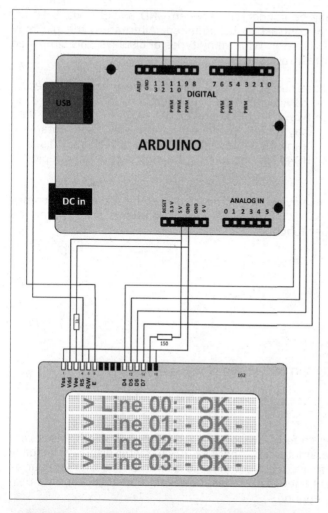

Figure 6.9:
4-Line, 16-Character LCD Connected to the Arduino

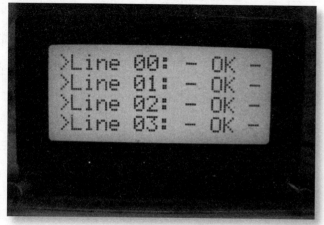

Figure 6.10:
LCD in Action

Listing 6-7: *LCD Display Test*

```
// Listing 6-7
// LCD Display Test

#include <LiquidCrystal.h>

// create lcd object, specifying pins
LiquidCrystal lcd(12, 11, 5, 4, 3, 2);

void setup() {
  // initialize lcd object for 16 columns, 4 rows
  lcd.begin(16, 4);
  delay(1000); // reset delay for slower LCDs
}

void loop() {
  // set the cursor to column 0, row 0
  lcd.setCursor(0, 0);
  lcd.print(">Line 00: - OK -");

  lcd.setCursor(0, 1);
  lcd.print(">Line 01: - OK -");

  lcd.setCursor(16, 0);
  lcd.print(">Line 02: - OK -");

  lcd.setCursor(16, 1);
  lcd.print(">Line 03: - OK -");
}
```

6 DISPLAYS AND DISPLAY TECHNIQUES

7
Measurement and Sensors

Sensors are to machines what senses are to humans. Almost all animal senses can be simulated using technology. Cameras, and in simple cases, photodiodes, are electronic eyes. Sounds are detectable by microphones and other sound transducers at sensitivities similar to the human ear. Even temperature sense and touch can be replicated by sensors. Electronics offers further senses as well. Things that humans can't detect, such as electrical currents and voltages, even radioactivity and the presence of odorless, poisonous gases, are easy for devices to detect.

7.1 Flexible and Easy to Read: An LED Voltmeter

The Arduino makes it possible, with a few additional components, to put together a practicable voltage meter very easily. With just a few 5 V LEDs, a bar graph voltmeter may be realized. This type of display may be familiar from the level meters in hi-fi systems. In the automobile industry, perhaps for digital tachometers, this type of display is often used, as it allows a readable display of rapidly varying measurements. The length of a light bar is much easier to determine than a number on a conventional digital display, especially when the numbers are changing rapidly.

A bar graph display circuit is pictured in Figure 7.1. Note that, in this version, every LED element requires its own resistor (220 Ω). For test purposes, it is also possible to use only a single series resistor to ground, as in the POV circuit. This would be a perfect demonstration of how each LED dims as more LEDs are activated.

In the program, the Arduino ports in use are stored in an array. That way, all of the ports may be quickly addressed within a `for` loop. In `setup()`, the same ports are set as outputs.

The ADC value is stored in the `value` variable. The value in `delta` determines the display's step size. With `delta` = 50, a voltage of

$$U_{delta} = (delta / 1023)*U_{ref} = (50 / 1023)*5 \text{ V} = 0.24 \text{ V}$$

per LED element is the result. All LEDs will be lit from 9 x 0.25 V, or 2.25 V ('full-scale'), and up. Using the `delta` variable, the voltmeter may be very easily reconfigured.

In the main loop, the number of LEDs turned on is dependent on the ADC value. This first LED is always on, as a power indicator. The rest of the LEDs follow, depending on the measured voltage. After a short delay, the LEDs are turned off again. In this way, a dynamic bar whose length is proportional to the ADC input voltage is created.

7 MEASUREMENT AND SENSORS

Note that the input voltage should never exceed 5 V, as this could destroy the microcontroller. For higher voltages, one may use a voltage divider. Further details on this may be found in the next section.

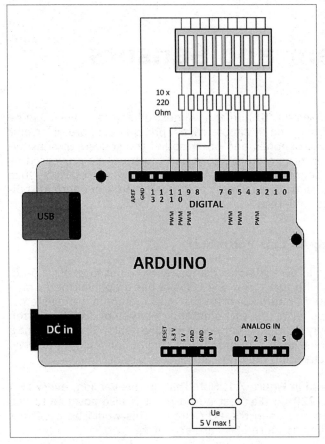

Figure 7.1:
Bar Graph Voltmeter Circuit Diagram

Listing 7-1: Bargraph Voltmeter

```
// Listing 0701
// Bargraph Voltmeter

const byte ledPin[] = {2, 3, 4, 5, 6, 7, 8, 9, 10, 11};
const int delta = 50;

int value;

void setup() {
  // all LED pins as outputs
  for (int x = 0; x < 10; x++) {
    pinMode(ledPin[x], OUTPUT);
  }
```

```
}

void loop() {
  value = analogRead(0);
  digitalWrite(ledPin[0], HIGH); // first LED always on

  // activate relevant LEDs
  for (int i = 1; i <= 9; i++) {
    if (value >= (i * delta)) {
      digitalWrite(ledPin[i], HIGH);
    }
  }
  delay(10);

  // all LEDs off
  for (int x = 0; x < 10; x++) {
    digitalWrite(ledPin[x], LOW);
  }
}
```

7.2 Volt / Ammeter: Precise Instrument for the Hobbyist's Lab

For precise measurement, one may resort to the classic 7-segment digit display. However, combinations of bar graph and digit display are possible. A number of digital multimeters work this way. Since we've already demonstrated the bar graph display in the previous section, we'll describe a voltmeter with an LC display here.

The basic functionality of a liquid crystal row display was explained in Section 6.7. Now, we need only determine the voltage value from the ADC input and send it to the display.

The Arduino's analog inputs are perfectly suited for the measurement of voltages. The only drawback is that they are only capable of directly measuring voltages of up to 5 V. When using the controller's internal reference voltage, this is reduced even further, to 1.1 V. However, this is not a serious problem, as the measuring range is easily extended as needed, using a simple voltage divider.

The use of the 1.1 V reference is always recommended when measuring external voltages. Using the power supply as a reference won't provide very accurate results, as the supply voltage may vary within relatively wide limits. Possible values range from 4.5 V to 5.5 V. In fact, some USB ports deliver even less than 4.5 V. From this alone, one may already reckon with accuracy errors of around 10%.

However, using the supply voltage has its advantages. There are, for example, sensors whose output values are proportional not only to the measured values, but also to the supply voltage. Should sensors such as these be fed with the same supply voltage as the microcontroller, the supply voltage will have no influence on the measured voltage.

Because we're interested in measuring external voltage sources, we will be using the internal reference. This is done with the function

7 MEASUREMENT AND SENSORS

```
analogReference(INTERNAL);
```

This will provide a relatively stable reference. According to the microcontroller's datasheet, the internal reference voltage is between 1.1 V and 1.2 V. Although this is still a ±10% margin of error, the voltage will be very stable and will hardly vary at all with variations in the supply voltage or ambient temperature. If even more precision is required, an external reference voltage may be used. This is connected to the AREF pin, and would be indicated using

```
analogReference(EXTERNAL);
```

Due to the ATmega328's variation in internal reference voltages, it is necessary to calibrate any measuring devices that are designed around it. This will be discussed in more detail shortly.

For measurements, the input voltages must be reduced to appropriate values. Figure 7.2 shows a suitable circuit. A voltage divider consisting of one 22 kΩ and one 1 MΩ resistor extends the limit of measurement to 51.1 V.

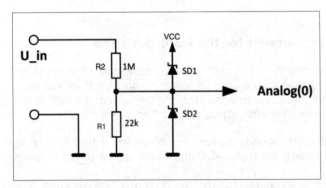

Figure 7.2:
Using a Voltage Divider to Extend the Voltmeter's Range of Measurement

Voltages higher than 50 V are rarely used in electronic circuits. In addition, one may only work on such voltages with safety-approved instruments. Heed this:

Note
    Voltages above 50 V may be lethal in some circumstances.

The multiplier for the given resistors is calculated as

$$V_a = R1 / (R1 + R2) = 22 \text{ k}\Omega / 1022 \text{ k}\Omega \approx 0.0215$$

So, 50 V will be reduced to 1.075 V when multipled by 0.0215 in this manner and may then be measured against the internal reference voltage.

The voltage divider's influence must naturally be accounted for in software. In our sketch, the calibration factor is the variable `cal`. This is one factor in the example sketch, but there is another constant, 0.94, which takes into account tolerances for the reference voltage source and the resistance values.

For additional overload protection, Schottky diodes may be employed as illustrated in Figure 7.2. SD1 protects against excessive voltages, while SD2 protects against reversing the input voltage polarity.

Listing 7-2: *LCD Voltmeter*

```
// Listing 7-2
// LCD Voltmeter

#include <LiquidCrystal.h>

const int adcChannel = 0;
const float vRef = 1.10;                    // internal reference voltage
const float r1 = 1000;                      // for voltage divider
const float r2 = 22;                        // for voltage divide
const float cal = 0.94 * (r1 + r2) / r2;    // calibration factor

int dac0;

// create lcd object, specifying pin numbers
LiquidCrystal lcd(12, 11, 5, 4, 3, 2);

void setup() {
  lcd.begin(16, 4);
  delay(1000); // delay for slower LCD modules
  // print start message
  lcd.print("Voltmeter");
  delay(1000);
  analogReference(INTERNAL);
}

void loop() {
  dac0 = analogRead(adcChannel);
  lcd.setCursor(0, 1);
  lcd.print(dac0 * vRef / 1023 * cal);
  lcd.print("V");
  delay(1000);
  lcd.setCursor(0, 1);
  lcd.print("            ");
}
```

With the help of a reference voltmeter, extremely high accuracy may be achieved. The following table shows a comparison between values measured with a calibrated desktop voltmeter and the Arduino voltmeter circuit.

7 MEASUREMENT AND SENSORS

Reference Measurement (V)	Arduino Measurement (V)
0.201	0.19
0.501	0.48
1.009	1.00
2.000	1.99
3.006	3.01
5.010	5.02
10.080	10.10
15.010	15.03
20.020	20.05
25.010	25.07
30.080	30.10

Table 10: Comparison between Reference and Measured Values

Calibration of the voltmeter using the `cal` factor is known as software calibration. Hardware calibration may also be used. For this, a potentiometer must be added to the circuit. A trimmer pot is especially suited for this, as it allows for very precise adjustment, and is also characteristically very stable over time.

Again, we compare with a calibrated reference voltmeter. This time, however, no calibration factor is set in the program, but rather the trimmer potentiometer is adjusted until the Arduino voltmeter displays the desired value.

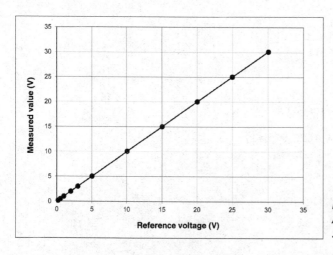

Figure 7.3: Arduino Voltmeter Calibration Slope

The advantage of this method is that the board need not be reprogrammed every time additional calibration is required. Rather, should operation be required in extreme conditions (e.g. very high or very low temperatures), a simple recalibration may be done using the trimmer pot.

The disadvantage is that an additional hardware component is required. Also, potentiometers, being electromechanical components, always exhibit a certain amount of drift, i.e. the electrical characteristics are affected by environmental factors, so the long-term stability of the calibration is reduced.

The selected potentiometer has a maximum resistance of 1 kΩ. This allows the measured value to be adjusted by about 5%. Larger potentiometers will allow wider adjustment ranges, but with reduced precision.

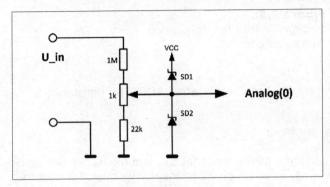

Figure 7.4:
Voltage Divider with Trimmer Potentiometer for Adjustment

Extensions and excercise

- Combine the bar graph display circuit with the LCD circuit, so that a precise digital value is visible next to a fast-changing rough measure of the voltage.

- In circuit development, it's often useful to measure several voltages simultaneously. There are still free ADC channels available on the Arduino. Expand the simple voltmeter to a multiple-voltage meter.

For an ammeter, a 1.0 Ω resistor may simply be connected to the voltmeter input, over which a 1 A current will deliver a voltage drop of 1.0 V. Even better, use a resistance of 0.1 Ω and then select a suitably dimensioned voltage divider.

7.3 Kiloohmmeter for Specific Applications

Aside from voltage and current, the Arduino may also be used to measure resistance. In principle, a voltage divider consisting of measurement resistor R_m and device-under-test R_x (see Figure 7.5).

7 MEASUREMENT AND SENSORS

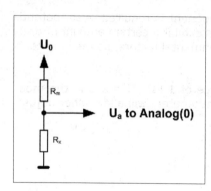

Figure 7.5:
Voltage Divider for Resistance Measurement

The output voltage V_a is determined by the Arduino's internal ADC. The resistance is then determined by

$$R_x = V_{out} * R_m / (V_{in} - V_{out})$$

The mathematically oriented user will immediately see that this is a measurement function of the form $y(x) = x * k_1 / (k_2 - x)$. This is highly nonlinear. The basic form of this graph is shown in Figure 7.6.

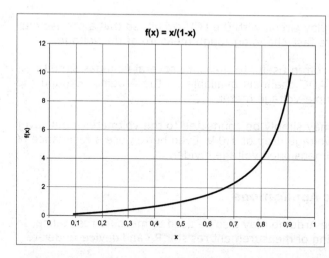

Figure 7.6:
Voltage Divider Measurement Function

For very small resistance values, the curve is quite flat, so even large changes in resistor values result in small voltage changes, which means significant measurement errors. The same applies to very large resistances, which occupy a very steep part of the curve. Thus, the best accuracy is achieved when the resistor values of R_m and R_x are roughly equal. Naturally, it's possible to adjust the measurement range across the series resistor. At R_m = 10 kΩ, values of between around 100 Ω and 100 kΩ may be measured quite accurately. For increased accuracy, measurement resistor R_m may be replaced. Although commercial ohmmeters work using a somewhat different method, they also have different range settings in order to offer improved accuracy.

In the following sketch, the output voltage of the voltage divider is measured and resistance value R_x is calculated according to the formula above. If the resistor ratio leads to reduced measurement accuracy, a warning is given. The measurement output may be displayed in the Serial Monitor. For a stand-alone device, an LCD may be used instead.

Listing 7-3: *Kiloohmmeter*

```
// Listing 7-3
// Kiloohmmeter

const int adc0 = 0;
const float r = 10.0; // reference resistor in kOhm

float x = 0.0;
float rX = 0.0; // resistance of DUT

void setup() {
  Serial.begin(9600);
}

void loop() {
  x = analogRead(adc0);
  if ((x > 10) && (x < 950)) {
    rX = x / (1023 - x) * r;
    Serial.print("R = ");
    Serial.print(rX);
    Serial.println(" kOhm");
  }
  if (x <= 10) {
    Serial.println("underflow");
  }
  if (x >= 950) {
    Serial.println("overflow");
  }
  delay(500);
}
```

7.4 No More Faulty Electrolytic Capacitor Woes: The 'Elcaduino' Tester

Electrolytic capacitors are among the most troublesome of electronic components. While they are essential to power supplies of all kinds due to their high capacitances, they have a high rate of failure due to their internal construction. The main reason for the failure of an electrolytic capacitor is the drying out of its electrolyte. This leads to a considerable reduction in capacity over time, and this problem is a common reason for failures in power supplies, PCs and audio amplifiers.

The device described here makes it possible to determine the capacitance of an electrolytic capacitor with great accuracy.

7 MEASUREMENT AND SENSORS

Caution
Before connecting a capacitor to the device, ensure that the capacitor is completely discharged, as the energy discharge from a capacitor may destroy the Arduino's input stage!

The measurement principle is simple: First, the electrolytic capacitor's positive terminal is switched to LOW potential via the Arduino pin, to create a defined initial state. Thereafter, this pin is changed to an input pin, i.e. high-impedance. Now, using the function

```
digitalWrite(activePin, HIGH);
```

the Arduino pin's internal pull-up resistor is activated, and the capacitor charges via this resistor. The sketch then measures how long it takes for the capacitor to charge up to about half of the supply voltage. This time is a direct measure of the capacitance.

$$T_{charge} = R * C_x$$

where R is the internal pull-up resistance—approximately 60 kΩ. The charging time is converted directly into a capacitance value in nF using calibration constant `cal`. If the calculated value is greater than 1,000, it is automatically displayed in µF. So, this simple device is capable of measuring not only electrolytic capacitors in the microfarad range, but also small non-polarized capacitors of down to about 10 nF. With even smaller capacitances, the required measurement times are too small and an alternative measurement method must be devised, as discussed in the following section.

Again, the measurement output may be displayed in the Serial Monitor.

In the program, first the calibration constant, `cal`, is defined. By measuring known capacitances, this constant may be adjusted for further precision. Then follows the now-familiar LCD initialization.

In `setup()`, a short startup message is sent to the LCD. Then, in the `main()` loop, the measurement procedure described above is carried out. Finally, the measured capacitance is output via the auto-ranging function.

Listing 7-4: *Elcaduino*

```
// Listing 7-4
// Elcaduino

// capacitance meter for 1 nF to 100 uF
// with LCD display

#include <LiquidCrystal.h>

const float cal = 0.5;              // calibration constant
const int activePin = 12;

float chargingTime;
float capacitance;
```

```
// initialize the LCD interface pins
LiquidCrystal lcd(8, 9, 5, 4, 3, 2);

void setup() {
  // initialize LCD object for 16 columns, 4 rows
  lcd.begin(16, 4);
  delay(1000); // delay for slower LCD modules
  lcd.setCursor(0, 0);
  lcd.print("CapMeter nF - uF");
  delay(999);
  lcd.setCursor(0, 0);
  lcd.print("                ");
}

void loop() {
  // discharge cap
  pinMode(activePin, OUTPUT);
  digitalWrite(activePin, LOW);
  chargingTime = 0.0;
  delay(1000);

  // charge cap
  pinMode(activePin, INPUT);
  digitalWrite(activePin, HIGH);

  // measure
  while (!digitalRead(activePin)) {
    chargingTime++;
  }
  capacitance = chargingTime * cal;         // calibration

  if (capacitance < 1000) { // ranging
    lcd.setCursor(3, 0);
    lcd.print(capacitance);
    lcd.print(" nF ");
  }
  else {
    capacitance /= 1000;
    lcd.setCursor(3, 0);
    lcd.print(capacitance);
    lcd.print(" uF ");
  }
  delay(1000);
}
```

7.5 'Picofaraduino': Measuring Smaller Capacitances

The measurement method from the previous section is not suitable for capacitances of less than about 10 nF, as the measurement speed required is too high.

7 MEASUREMENT AND SENSORS

For smaller capacitors, in the picofarad range, the method must be modified. To achieve measurable charge times, a very high-value resistor is required. A 10 MΩ resistor is ideal for this. Should this be hard to find, two 4.7 MΩ resistors in series will work, or even up to ten 1 MΩ resistors in series. The calibration constant, `cal`, will have to be set accordingly.

In this sketch, the CapSense library is used, so it must be installed. CapSense.h is available for download from

> github.com/GoogleChrome/chrome-app-samples/tree /master/serial/adkjs/firmware/arduino_libs/CapSense

The sketch itself is very simple. After setting the measurement and sensor pins, the result may be determined using a library function. After the necessary compensation, the result of the measurement is output via the serial interface.

Note
This program must be compiled using Arduino IDE version 0022.

Listing 7-5: *Picofaraduino*

```
// Listing 7-5
// Picofaraduino

#include <CapSense.h>

const long cal = 155.0; // calibration factor

// 10 megohm between pins 7 & 11
// unknown capacitor between pin 11 and GND
CapSense cs_7_11 = CapSense(7, 11);

void setup() {
  Serial.begin(9600);
}

void loop() {
  long v = cs_7_11.capSense(30);
  Serial.println(v / cal); // send sensor output to PC
  delay(1000);             // arbitrary delay to limit serial data
}
```

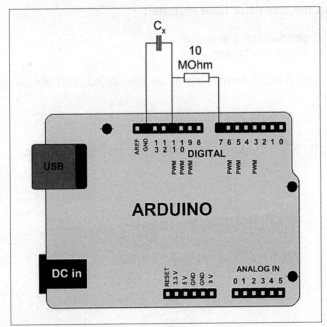

Figure 7.7:
Picofaraduino

Exercise
- How may both the Elcaduino and the Picofaraduino be integrated into a single device?

7.6 'Transistino': Transistor Tester

Passive components are not the only ones that may be measured using such simple means. Active components such as transistors may likewise be evaluated. The circuit shown in Figure 7.8 measures the main characteristics of a transistor.

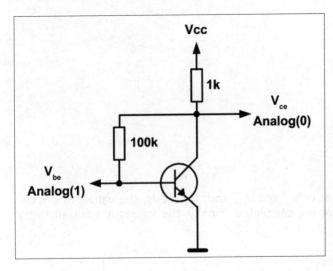

Figure 7.8:
Transistor Tester

7 MEASUREMENT AND SENSORS

The base voltage, V_{be} tells us about the transistor base material:

$$V_{be} \approx 0.3 \text{ V: germanium transistor}$$
$$V_{be} \approx 0.7 \text{ V: silicon transistor}$$

From the collector voltage, V_{ce}, the collector current, I_c, may be calculated, and similarly the base current from V_{be}. Then

$$\beta = I_c / I_b$$

is the transistor's DC current gain, β. Typical values for small signal transistors are in the range of 100 to about 800.

Listing 7-6: *Transistino*

```
// Listing 7-6
// Transistino

float vCe, vBe, beta;

void setup() {
  Serial.begin(9600);
}

void loop() {
  vCe = 5.0 * analogRead(0) / 1023;
  vBe = 5.0 * analogRead(1) / 1023;
  beta = (5 - vCe) / (vCe - vBe) * 100;

  Serial.print("Vbe = ");
  Serial.println(vBe);
  Serial.print("Ib = ");
  Serial.print((vCe - vBe) * 10);
  Serial.println(" uA");
  Serial.print("Ic = ");
  Serial.print(5 - vCe);
  Serial.println(" mA");
  Serial.print("beta = ");
  Serial.print(beta);
  Serial.println(" @");
  Serial.println();
  delay(1000);
}
```

The program determines the values of V_{ce} and V_{be}, and, from this, the values of the collector current and the base current are calculated. Finally, the values are output to the serial interface.

The program is very useful if you want to test a large number of transistors. From this, one can get not only the information on whether a special transistor works or not, but also the value of its current gain. So, with this sketch, it's possible to sort transistors of the same type by their current gain. In this way, one could select 'matched pairs' for current mirrors and similar applications.

Extension and exercise

- The circuit works for NPN transistors. Develop a counterpart for PNP types.

7.7 A Simple NTC Thermometer

NTC thermistors, temperature-dependent resistors with negative temperature coefficients, are an easy way to measure temperatures. The resistance of an NTC element decreases with increasing temperature. A major disadvantage of the NTC is its non-linear characteristic (see Figure 7.9).

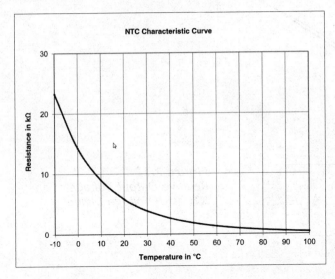

Figure 7.9:
NTC Characteristic Curve

The situation is somewhat aggravated when using an NTC in a voltage divider circuit. As we know from the section on resistance measurement, the relationship between the resistance values in a voltage divider and the corresponding output voltage is not linear. An appropriate circuit can compensate for these two nonlinear effects to a certain extent (see Figure 7.10), but there will always be a slight deviation from the ideal line.

For the fixed resistance in the voltage divider, it is best to select the same value that the NTC exhibits at 25 °C (the 'face value' - in this example, an NTC type with a 4.7 kΩ face value is used).

7 MEASUREMENT AND SENSORS

In the sketch, linear approximation is used. For this, the parameters `offset` and `cal` are created. From the calibration curve in Figure 7.10, a slope of

$$S = (0.78 - 0.25) / 60\ °C = 0.009\ /\ °C$$

is derived, as well as an offset of

$$d = 0.25$$

With the maximum ADC value being 1,023, the calibration formula is

$$T = 0.11 * \mathtt{ADC\_value} + 28$$

Thus, `offset = 28` and `cal = 0.1`. With these two values, the thermometer should deliver reasonably accurate results.

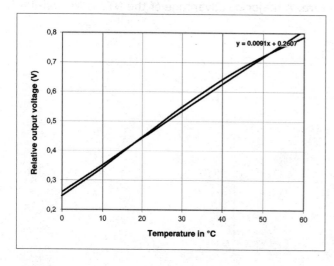

Figure 7.10:
Relative Output Voltage of an NTC with Voltage Divider

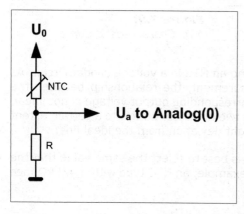

Figure 7.11:
NTC with voltage divider

Listing 7-7: *NTC Thermometer*

```
// Listing 7-7
// NTC Thermometer

const int offset = 28;    // offset for NTC 4k7
const float cal = 0.11;   // calibration factor

void setup() {
  Serial.begin(9600);
}

void loop() {
  int adcValue = analogRead(0);
  float temp = adcValue * cal - offset;
  Serial.print("Temp = ");
  Serial.print(temp);
  Serial.println(" C");
  delay(1000);
}
```

Significantly better accuracy is achieved by calibrating the measurements against a calibrated thermometer. For this, it's best to compare measurements at a few temperature points and calculate the optimal values for `offset` and `cal`.

The calculated values are output via the serial interface and viewed in the Serial Monitor.

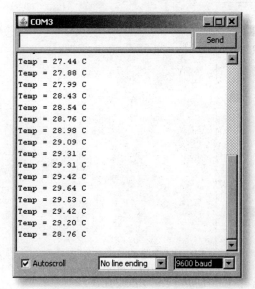

Figure 7.12:
Temperature Output to the Serial Monitor

7.8 Hot or Cold? Temperature Measurement Using the AD22100

The AD22100 is a lot more precise than a simple NTC. With this temperature sensor in a transistor housing, we have a complete, integrated measurement system.

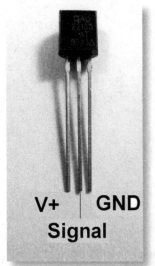

Figure 7.13:
AD22100 Temperature Sensor

Connecting the sensor to Arduino is simple. The following figure shows the corresponding schematic.

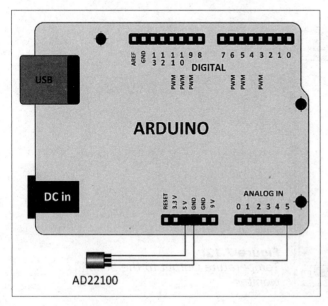

Figure 7.14:
Temperature Measurement Using the AD22100

Instead of a curve, this device provides a strongly linear relationship between temperature and output voltage:

$$V_{out} = 1.375 \text{ V} + 0.0225 \text{ V} / \text{°C} * \delta$$

This eliminates the need for time-consuming calibration. The sketch is equally as simple:

Listing 7-8: *AD22100 Thermometer*

```
// Listing 7-8
// AD22100 Thermometer

int adcResult;
float v;
float temp;

void setup() {
  Serial.begin(9600);
}

void loop() {
  adcResult = analogRead(5);
  v = adcResult * 5.0 / 1023; // calculate sensor voltage, 10 bit & Uref=5V
  temp = ((v - 1.375) / 0.0225); // calibrate AD22100
  Serial.println(temp); // send result to PC
  delay(1000);
}
```

After the ADC value is read, it is converted into a voltage value. The temperature is then easily calculated using the above-mentioned transformation. Finally, the temperature is transmitted, via the serial interface, to the PC.

7.9 Remote Thermometer

A big advantage of temperature sensors is their ability to send temperature measurements from remote or hard-to-reach places. With an ordinary mercury thermometer, one is only able to measure the temperature in the immediate area of the thermometer. For temperature monitoring within an enclosure, for example in a PC, such a thermometer is unsuitable, as it would require the case to be open in order to see the temperature.

With an electronic thermometer, on the other hand, the measurement location may be far from the display unit, or even in a difficult-to-reach place. This makes it easy to measure, for example, the temperature in an engine compartment, on the axis of a turbine, or in a closed electrical cabinet, and easily display it on a digital display unit inside the vehicle or in a control center.

7 MEASUREMENT AND SENSORS

The AD22100 is characterized by its small transistor-sized package. If a shielded cable is soldered to the terminals, one may easily bridge distances of several meters between the location of measurement and the display. Even a simple, unshielded three-wire ribbon cable may transmit measurements over smaller distances. In this way, for example, the temperature in a refrigerator or freezer may be monitored.

The CPU temperature in a computer may also be measured. This method is often used in PC tuning, especially when the processor is overclocked and the user is concerned about thermal overload. One may also measure the outdoor temperature and monitor it from inside one's home. A very thin ribbon cable is used for this, which may be run between the window and the window frame. There are special thin films with printed conductors available for this purpose. This ensures that the door or window seals are not damaged.

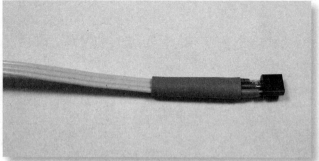

Figure 7.15:
Temperature Sensor with Ribbon Cable

7.10 'Thermodino': Precision Thermometer with 7-Segment Display

When a complete, self-contained thermometer is required (one that works without a PC on hand to display the temperature), we turn again to the various display possibilities available to the Arduino. One is shown in Figure 7.16. In addition to the sensor, a connected 4-digit, 7-segment display is shown.

For a 4-digit display to be feasible for temperature display, a minus sign must be added for negative readings. If need be, this may simply be a rectangular LED. Control of the minus sign is via an additional I/O pin, Digital Pin 8. For this, only the sign of the calculated temperature need be established. The numerical digits only display the absolute value of the temperature.

'THERMODINO': PRECISION THERMOMETER WITH 7-SEGMENT DISPLAY   7.10

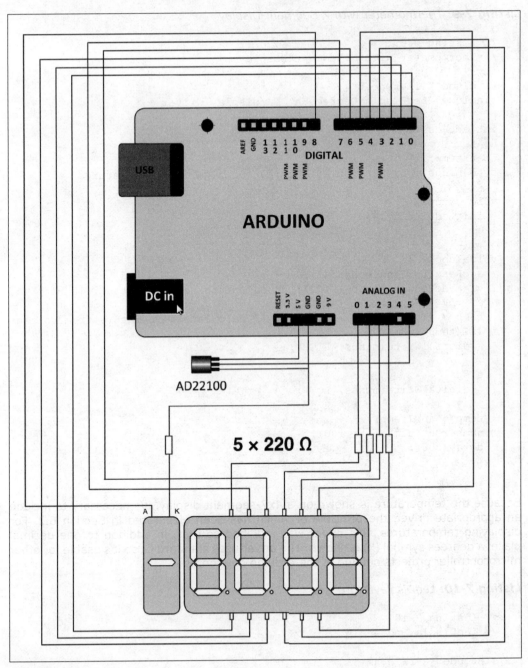

Figure 7.16:
Digital Thermometer with 7-Segment Display

7 MEASUREMENT AND SENSORS

Listing 7-9: *Thermometer with 7-Segment Display*

```
// Listing 0709
// Thermometer with 7-Segment Display

#include "TimerOne.h"
#include "LedDisplayDegrees.h"   //display driver

int adcResult;
float v;
float temp;
byte minus = 8; // Display ‚Äù-‚Äù for negative temperatures

void setup() {
  initLedDisplay();
}

void loop() {
  adcResult = analogRead(5);
  v = adcResult * 5.0 / 1023; // calculate sensor voltage, 10 bit & Vref =5V
  temp = ((v - 1.375) / 0.0225); // calculate temperature from voltage
  if (temp < 0) {
    digitalWrite(minus, HIGH);
  }
  else {
    digitalWrite(minus, LOW);
  }
  temp = abs(temp);
  numberOutput(temp * 100);
  delay(300);
}
```

Because the temperature is shown on a 4x7-segment display, it's necessary to include an appropriate driver, the principle of which has been elaborated in Section 6.2. For displaying temperatures, this driver was modified so that, in addition to one decimal place, a degrees symbol (°) is shown. The driver uses standard C, so it's usable for other microcontroller projects outside of the Arduino environment.

Listing 7-10: *LedDisplayDegrees.h*

```
// Listing 7-10
// LedDisplayDegrees.h

#include <avr/io.h>
#include <avr/interrupt.h>
#include <util/delay.h>

// 7-segment codes
const int numbers[11][8] = {
```

```c
  // e, d, DP, c, g, b, f, a
  {0, 0, 1, 0, 1, 0, 0, 0}, // 0
  {1, 1, 1, 0, 1, 0, 1, 1}, // 1
  {0, 0, 1, 1, 0, 0, 1, 0}, // 2
  {1, 0, 1, 0, 0, 0, 1, 0}, // 3
  {1, 1, 1, 0, 0, 0, 0, 1}, // 4
  {1, 0, 1, 0, 0, 1, 0, 0}, // 5
  {0, 0, 1, 0, 0, 1, 0, 0}, // 6
  {1, 1, 1, 0, 1, 0, 1, 0}, // 7
  {0, 0, 1, 0, 0, 0, 0, 0}, // 8
  {1, 0, 1, 0, 0, 0, 0, 0}, // 9
  {1, 1, 1, 1, 0, 0, 0, 0}  // degree
};

// store single digits
volatile int D1, D2, D3, D4;
volatile uint8_t activeDigit = 0;

#define sbi(PORT, bit) (PORT |= (1 << bit))      // set bit in PORT
#define cbi(PORT, bit) (PORT &= ~(1 << bit))     // clear bit in PORT

// Write number at indicated position to display
void digit(int value, int pin) {
  sbi(PORTC, pin); // pin high => current digit on
  // cbi(PORTC, pin); // enable for common cathode

  PORTD = 0b11111111;
  // PORTD = 0b00000000; // enable for common cathode

  // activate one segment after another
  for(int i = 0; i <= 7; i++) {
    if (0 == numbers[value][i]) {
      cbi(PORTD, (i)); // set segment
      // sbi(PORTD, (i)); // enable for common cathode
      _delay_ms(1);
      sbi(PORTD, (i)); // clear segment
      // cbi(PORTD, (i)); // enable for common cathode
    }
  }
}

// distribute number to digits
void numberOutput(uint16_t number) {
  D1 = number / 1000; number %= 1000;
  D2 = number / 100;  number %= 100;
  D3 = number / 10;   number %= 10;
  D4 = number;
}
```

7 MEASUREMENT AND SENSORS

```c
void setLowDot() { // draw decimal point
  PORTD = 0b11111011;
  // PORTD = 0b00000100; // enable for common cathode
  sbi(PORTC, PORTC0);
  // cbi(PORTC, PORTC1); // enable for common cathode
}

void setDegree() { // draw degree symbol
  sbi(PORTC, PORTC3); // pin high => current digit on
  // cbi(PORTC, PORTC3); // enable for common cathode
  PORTD = 0b11111111;
  // PORTD = 0b00000000; // enable for common cathode

  // activate one segment after another
  for (int i = 0; i <= 7; i++) {
    if (0 == numbers[10][i]) {
      cbi(PORTD, (i)); // set segment
      // sbi(PORTD, (i)); // enable for common cathode
      _delay_ms(1);
      sbi(PORTD, (i)); // clear segment
      // cbi(PORTD, (i)); // enable for common cathode
    }
  }
}

void updateDisplay() {
  PORTC = 0b00000000; // all digits off
  // PORTC = 0b00001111; // enable for common cathode
  if (0 == activeDigit) {
    setDegree();
  }
  if (1 == activeDigit) {
    digit(D1, PORTC0);
  }
  if (2 == activeDigit) {
    digit(D2, PORTC1);
  }
  if (3 == activeDigit) {
    digit(D3, PORTC2);
  }
  if (4 == activeDigit) {
    setLowDot();
  }
  activeDigit ++;
  if (5 == activeDigit) {
    activeDigit = 0;
  }
}

void initLedDisplay() {
  DDRD = 0b11111111;
```

```
DDRC = 0b00001111;
Timer1.initialize(3000); // Interrupt every 3000 us = 3 ms
Timer1.attachInterrupt(updateDisplay);
}
```

Extensions and exercise

Expand this idea to
- a frost warning indicator that puts out a warning signal for temperatures under 3 °C

- a freezer warning device, which should activate an alarm for temperatures above -10 °C

7.11 When Are We Most Comfortable? – The Hygrometer

After temperature, relative humidity is the second-most important parameter when it comes to comfort. As such, dry cold is usually perceived as much more pleasant than humid cold weather. Various plant and animal species require, in addition to high temperatures, high humidity. In technology, we are usually interested in low humidity, as a relatively high water content in the air could easily lead to corrosion damage and the like. Of special importance is the so-called dew point. When humidity reaches a certain level, water may condense on surfaces, resulting in the known negative effects for electronic circuits. For this reason, in server rooms, humidity as well as temperature is measured and automatically controlled.

Humidity sensors are technically more complicated than temperature sensors. Sometimes, this is clearly reflected in the price. Capacitive transducers for determining relative humidity are particularly expensive. Hydro-resistive sensors such as the SHS A2 are somewhat cheaper, but, for the price, one achieves a lower precision. For a simple living room hygrometer, however, the SHS A2 is accurate enough.

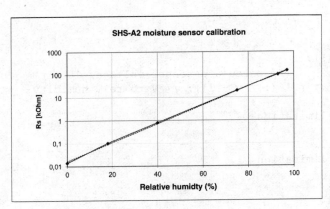

Figure 7.17:
Humidity Sensor Calibration

The following logarithmic transfer function can be derived from the calibration of an SHS A2 humidity sensor:

7 MEASUREMENT AND SENSORS

$$rF\,(\%) = 10.6 * \ln(61 * Rs\,/\,k\Omega)$$

and, with the already known voltage divider function, one gets:

$$rF\,(\%) = 10.6 * \ln(61 * R * (1\,/\,V_a) - 1).$$

In the program, the two calibration factors are defined as variables:

```
float fact = 10.6;
float expCoef = 61;
```

These are easily modified for different sensor types.

This application is also a good example of the mathematical abilities of the microcontroller. Achieving a linear response from a sensor with an exponential output by analogue means alone would be significantly more complicated. With our microcontroller, this is achieved with a single line of code.

Listing 7-11: *SHS A2 Hygrometer*

```
// Listing 7-11
// Digital Hygrometer using SHSA2

#include "TimerOne.h"
#include "LedDisplayDegrees.h"  // display driver

int adcResult;
float v, rH;
float r = 10;
float fact = 10.6;
float expCoef = 61;

void setup() {
  initLedDisplay();
}

void loop() {
  adcResult = analogRead(5);
  v = adcResult / 1023.0;        // calculate sensor voltage, 10 bit
  rH = fact * log(expCoef * r * (1 / v - 1)); // V to rel. humidity
  if (rH >= 99.9) {
    rH = 99.9;  // range limitation
  }
  if (rH <= 0) {
    rH = 0;     // range limitation
  }
  numberOutput(100 * rH);
  delay(300);
}
```

7.12 'Battduino': Capacity Measurement for Rechargeable Batteries

The ability to recharge batteries is a welcome feature. Trying to operate mobile electronic devices on disposable batteries would usually be very uneconomical. New battery packs would have to be purchased often, and the old ones discarded. The operation of laptops with disposable batteries is virtually unthinkable.

The well-known problem with rechargeables is their relatively rapid aging. NiMH batteries, according to manufacturer specifications, may be recharged up to 1,000 times, but, in practice, they rarely achieve this number. It would be desirable to determine the battery's capacity in any event. This would prevent unpleasant surprises due to prematurely flat batteries.

This is where the Battduino comes in. It produces a precise result of how many milliamp-hours a rechargeable battery is capable of delivering. For this, the relevant battery is first fully charged in accordance with its specifications and then connected to the Battduino. Battery discharge begins. The discharge current is set by load resistors (see Figure 7.18). A total resistance of 10 Ω provides a discharge current of about 120 mA for a 1.2 V NiCd cell. The series-parallel connection of four single 5 W resistors balances the load. A generous safety margin should always be provided for here. An additional fuse, such as a 1 A Polyfuse, should also not be omitted.

The LCD module connections are the same as in Figure 6.9.

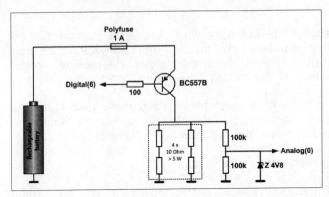

Figure 7.18:
Circuit Diagram for the Millamp-Hour Meter

When the Battduino is first turned on, it reports on operational readiness and cutoff current. The latter is set by the program variable i_min and may be changed to suit individual needs.

Since rechargeables should never be completely discharged, the Battduino disconnects the battery at a specified minimum current. With the load resistance of 10 Ω, this current corresponds with a minimum voltage of 2.5 V.

7 MEASUREMENT AND SENSORS

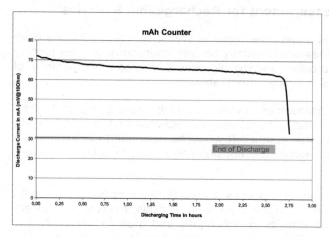

Figure 7.19:
Discharge Graph of a 180 mAh Battery

At 30-second intervals, the sketch calculates the charge drawn and sends this, along with the current consumption, to the serial interface. The LCD is updated once a second.

Two modes of operation are thus possible. In a stand-alone application, the device operates independently of a PC, updating the LCD once per second, then displaying the total mAh value after measurement is complete. In this case, the device may be powered by an external power supply.

If one wishes to capture the entire discharge curve, however, the Serial Monitor may be used to log the entire process. After completing the measurement, the data may then be evaluated in a spreadsheet program. The calculated discharge curve provides an excellent basis for assessing the condition of the battery (see Figure 7.19).

The ideal discharge curve is almost flat, with a steep termination at the end. A steeply sloping graph from the start usually indicates reduced capacity. When the discharge curve of a multi-cell battery reduces in steps, the individual cells have different capacities. In this case, the damaged cells should be replaced. There is a wealth of information available on the topic of evaluating rechargeable batteries.

By summing the measured values, the total capacity of the battery can also be derived from the discharge curve quite accurately.

'BATTDUINO': CAPACITY MEASUREMENT FOR RECHARGEABLE BATTERIES 7.12

Figure 7.20:
Proposed Battduino Construction in Enclosure

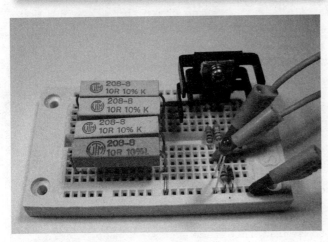

Figure 7.21:
Battduino Measurement Board

Listing 7-12: *Battduino*

```
// Listing 7-12
// Battduino

#include <LiquidCrystal.h>

const int adc0 = 0;                  // ADC channel 0 selected
const int curSwitch = 6;             // digital pin for current switch
const float calAdc = 2.2 * 5 / 10.23; // calibration constant for ADC to current conversion
                                     // R_L = 10 Ohm, 50% voltage divider
const int maxCtr = 30;               // print acc_i to RS232 every
```

7 MEASUREMENT AND SENSORS

```
"max_ctr" seconds
const int iMin = 250;              // minimum discharge current in mA

float i;      // current
float c;      // capacity
float accI;   // accumulated current
long ctr;     // counter for average
long adcVal;

// create lcd object, set pin numbers
LiquidCrystal lcd(12, 11, 5, 4, 3, 2);

void setup() {
  pinMode(curSwitch, OUTPUT);
  // initialize LCD number of columns and rows
  lcd.begin(16, 2);
  delay(1000); // delay for slower LCD modules
  Serial.begin(9600);

  // send startup string to LCD
  lcd.setCursor(0, 0); lcd.print("Battduino");
  delay(3000);
  lcd.setCursor(0, 0); lcd.print("               ");
  lcd.setCursor(0, 0); lcd.print("T=");
  lcd.setCursor(2, 0); lcd.print(maxCtr); lcd.print("s");
  lcd.setCursor(7, 0); lcd.print("M");
  lcd.setCursor(0, 1); lcd.print("in=");
  lcd.setCursor(3, 1); lcd.print(iMin); lcd.print("mA");
  delay(3000);
  lcd.setCursor(0, 0); lcd.print("   ");
  lcd.setCursor(0, 1); lcd.print("   ");

  // send startup string to serial interface
  Serial.println("Battduino");
  Serial.print("dT="); Serial.print(maxCtr); Serial.println("s");
  Serial.print("I_min="); Serial.print(iMin); Serial.println("mA");

  digitalWrite(curSwitch, LOW); // switch on discharge current LOW = on
}

void loop() {
  adcVal = analogRead(adc0);
  i = adcVal * calAdc;

  // dispaly current i
  lcd.setCursor(2, 1); lcd.print("       ");

  if (i < 10) {
    lcd.setCursor(5, 1); lcd.print(i);
```

```
  }
  if (i >= 10 & i < 100) {
    lcd.setCursor(4, 1); lcd.print(i);
  }
  if (i >= 100 & i < 1000) {
    lcd.setCursor(3, 1); lcd.print(i);
  }
  if (i >= 1000) {
    lcd.setCursor(2, 1); lcd.print(i);
  }
  lcd.setCursor(6, 1); lcd.print(" mA");

  // add up capacity
  c = c + i / 3600;

  // display total capacity
  {lcd.setCursor(0, 0); lcd.print("          ");}
  if (c < 10) {
    lcd.setCursor(3, 0); lcd.print(c);
  }
  if (c >= 10 & c < 100) {
    lcd.setCursor(2, 0); lcd.print(c);
  }
  if (c >= 100 & c < 1000) {
    lcd.setCursor(1, 0); lcd.print(c);
  }
  if (c >= 1000) {
    lcd.setCursor(0, 0); lcd.print(c);
  }
  lcd.setCursor(4, 0); lcd.print(" mAh");

  // calculate average current
  accI = (accI + i); ctr++;
  if (ctr >= maxCtr) {
    Serial.println(accI / maxCtr);
    // swich off current if minimum average current is reached
    if ((accI / maxCtr) <= iMin) {
      digitalWrite(curSwitch, HIGH);
      Serial.println("Minimum discharge current reached");
      while(1);
    }
    accI = 0; ctr = 0;
  }

  // wait one second and blink ":"
  lcd.setCursor(8, 0); lcd.print(":");
  delay(500);
  lcd.setCursor(8, 0); lcd.print(" ");
  delay(500);
}
```

7.13 Optical Sensors: Important for More than Just Photography

Apart from temperature sensors, light sensors are one of the most important components. What started out in the 70s as mysterious light barriers in elevators and at automatic doors has become standard safety equipment on vehicles and in buildings.

The first popular optical sensors were light-dependent resistors (LDRs). These were best suited for simple applications such as these light barriers and twilight switches. They were both very sensitive and very slow. They react in the millisecond range, and are thus unsuitable for high-speed applications such as signal or data transmission.

Photodiodes are orders of magnitude faster than LDRs. Today's common optical fiber transmission rates of several terabits per second are only possible with the use of optimized photodiodes.

If high speed is not a concern, the phototransistor is a good compromise. It is characterized by high sensitivity and a fast enough response. Its mechanical construction is much like that of an LED, making it cheap and widely available. It works similarly to an ordinary transistor. The difference is that, in a phototransistor, light shines directly onto the semiconductor through a transparent housing. There, charge carriers are released by the internal photoelectric effect. This photocurrent is amplified directly in the component, so a phototransistor is capable of switching small loads directly.

Phototransistors usually have only two external terminals - a collector and an emitter. There are some versions with an accessible base terminal that enables, for example, adjustment of the switching point. If the base is unconnected, it takes relatively long until the base-emitter zone is free of charge carriers. The phototransistor's slow turn-off characteristic is due to this.

On the other hand, phototransistors are much more sensitive than photodiodes. Applications may be found in light barriers, twilight switches and opto-isolators. For receiving remote control signals, however, photodiodes are usually used, as phototransistors are too slow. Remote controls, and often light barriers and opto-isolators, don't use visible light, but rather infrared, as both phototransistors and photodiodes have a higher sensitivity in this range.

The wavelength of maximum sensitivity in a silicon phototransistor is about 850 nm (near infrared) and falls off in the direction of shorter wavelengths (visible light, ultraviolet). The receiving wavelength range is limited by the silicon's band-edge energy to longer wavelengths of around 1,100 nm.

For measurement of light intensity, either photodiodes or phototransistors may be used. The faster switching time and more linear curve of the photodiode are contrasted with the significantly higher sensitivity of the phototransistor.

The BPW40 is a universal, widely available phototransistor. The most significant characteristics of the BPW40 are summarized in the following table.

Maximum voltage	25 V
Dark current	100 nA
Collector current at 1 mW/cm²	0.5mA
Maximum collector current	30 mA

Table 11: BPW40 Phototransistor Characteristics

A guide to the expected photocurrents in specified lighting conditions is shown in Table 12.

	Light Intensity	BPW40 Collector Current
Broad daylight	10,000 lx	20 mA
Office or room lighting	1,000 lx	2 mA
Typical hallway lighting	100 lx	0.2 mA
Absolute darkness	0 lx	100 nA

Table 12: Typical Illumination Examples

Additional data may be found in the BPW40 datasheet.

In addition to the reflex light described in the next section, photodiodes and phototransistors may also be used as receivers for infrared remote control. In Section 11.2, photosensitive components play an important role in automatic control.

7.14 Reflex Light for Geocaching

Geocaching is a new type of game played by GPS users. It's basically about finding a 'treasure' in an open area. The geographical coordinates of the treasure are announced, usually online, and the treasure hunter must find the exact spot with the help of a GPS receiver.

In order to offer a reasonable chance of finding the treasure, even in twilight or dark conditions, a so-called reflex light may be employed. These electronic reflectors send out light signals when they detect light, e.g. from the beam of a flashlight.

Of course, a complete geocaching expedition need not always be arranged. A reflex light may also be hidden in one's own garden as a challenge for children or friends.

The circuit diagram for the reflex light is shown in Figure 7.22. If the phototransistor is illuminated by a flashlight, the voltage at analog input A0 rises. When the `threshold` variable exceeds a certain specified value, an LED connected digital pin 3 flashes three times.

The accompanying sketch is shown in Listing 7-13.

7 MEASUREMENT AND SENSORS

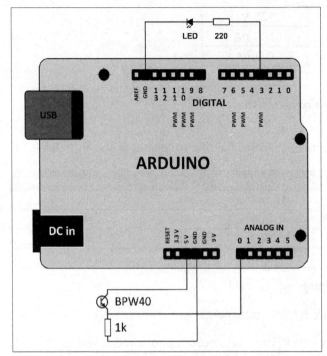

Figure 7.22: Reflex Flasher

Listing 7-13: Reflex Flasher

```
// Listing 7-13
// Reflex Flasher

const byte ledPin = 3;
const byte sensPin = 0;
const int threshold = 500;

void setup() {
  pinMode(ledPin, OUTPUT);
}

void loop() {
  delay(10);
  int lightIntensity = analogRead(sensPin);
  if (lightIntensity > threshold) {
    for(int i = 0; i < 3; i++) {
      analogWrite(ledPin, 255); delay(300);
      analogWrite(ledPin, 0); delay(300);
    }
  }
}
```

7.15 For Professional Photographers: A Digital Light Meter

The phototransistor is put to a more demanding task in this project. Not only light and dark needs to be differentiated, but the actual light intensity on the photo element must be measured as accurately as possible. Devices that detect brightness are called light meters, and their typical application is in the field of photography. Calculating optimal aperture and shutter speed values is much easier with knowledge of the light intensity. Illumination intensity is also important in the field of industrial safety, where the light intensity prescribed by professional insurance associations may be accurately verified using a light meter. The devices thus contribute directly to the improvement of safety in the workplace.

The circuit diagram for the light meter is shown in Figure 7.23. The phototransistor makes up half of a voltage divider, along with a 1 kΩ resistor. This will cover the most common light levels quite well. If particularly low light levels are to be measured, a 10 kΩ resistor may be used instead. The 10 µF capacitor is used to smooth the signal. Without it, especially in artificial light, the display would change rapidly, as the system would not be able to react quickly enough to the light and dark periods inherent in lamps driven by mains alternating current.

One may refer to Table 12 for calibration of the device. The values provided are, of course, rough guidelines. A more accurate calibration may be carried out with a reference light meter.

TimerOne.h is used again for the display driver. The display driver is slightly a modified version of that used in the thermometer, as a light meter requires neither a decimal point nor a °C symbol.

The calibration constant, `cal`, is set to `1` in the example sketch. The usual brightness ranges are thereby covered. Relative measurements are also possible with reasonable accuracy. The `cal` constant may be adjusted for an absolute calibration against a reference light meter.

In the main loop, the measured value is measured ten times consecutively and the results summed. When the parameter is divided by 10 when it is passed to the `numberOutput()` function, an average is calculated and displayed. Along with the capacitor, this contributes to a more stable display.

7 MEASUREMENT AND SENSORS

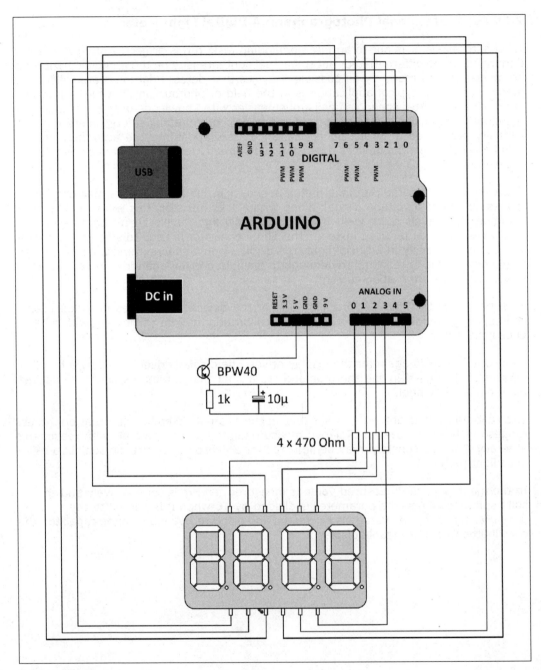

Figure 7.23:
Digital light meter

Listing 7-14: *Digital Luxmeter*

```
// Listing 7-14
// Digital Luxmeter

#include <TimerOne.h>
#include "ledDisplayV2.h" // display driver

const int cal = 1;
int adcResult;

void setup() {
  initLedDisplay();
}

void loop() {
  for (int i = 0; i < 10; i++) {
    adcResult += analogRead(5);
  }
  numberOutput(cal * adcResult / 10);
  adcResult = 0;
  delay(300);
}
```

7.16 Home 'Radar Station'': Distance Measurement Using Ultrasound

Ultrasound is very suitable for the medium-distance range of between a few centimeters and several meters. Its speed in air at 20 °C is around 343 m/s. A distance of 1 m is thus traversed in about 3 ms. This amount of time is easily and accurately measured by a microcontroller.

There are many application areas for such a measurement method. For example, a bridge's clearance height may easily be measured, or, by measuring a room's length and width, its area is easily calculated. Ultrasonic measurement systems have found wide use in robotics and in the automobile industry. Autonomous robot systems can detect obstacles, and distance from a wall may be reliably determined when parking a car.

In most of the above applications, ultrasonic transducers with a resonant frequency of about 40 kHz, which is easily generated using a microcontroller, are used. A full period of a single 40 kHz oscillation takes 0.025 ms, or 25 µs. The smallest unit of time measurable by a microcontroller with a 16 MHz clock is around 0.0625 µs, or 62.5 ns.

The oscillator frequency for an ultrasonic transmitter may thus be set to within a resolution of 0.1 kHz:

$$f+ = 1/(25 \text{ µs} - 62.5 \text{ ns}) = 40.1 \text{ kHz}$$
$$f0 = 1/25 \text{ µs} = 40.0 \text{ kHz}$$
$$f- = 1/(25 \text{ µs} + 62.5 \text{ ns}) = 39.9 \text{ kHz}$$

This is quite sufficient for tuning the controller to the resonant frequency of an ultrasonic capsule. The following code generates a pulse train with a frequency of 40 kHz. Such a train is ideal for ultrasonic radar. The `pulseLength` variable sets the number of oscillations in the pulse train:

```
for (int i = 0; i < pulseLength; i++) {
  PORTB = 0b00000100; _delay_us(12.5); // activate PB2 = Arduino D10
  PORTB = 0b00000000; _delay_us(12.5);
}
```

Why is a C `PORTB` command used here, instead of the more familiar Arduino function, `digitalWrite`(*pin, value*)?

The answer to this is due to one of the small disadvantages of the Arduino language. The `digitalWrite()` function consists of several C statements internally. The port to which the specified pin belongs must first be determined, requiring several instructions, which take time in the form of additional processor cycles. In comparison, the direct port command may be executed significantly faster.

Similarly, the Arduino `delayMicroseconds()` function comes with the following warning:

"This function works very accurately in the range 3 microseconds and up. We cannot assure that `delayMicroseconds()` will perform precisely for smaller delay times."

So, instead, the `_delay_us()` macro was used from the delay.h library of the C compiler. This works in the 1/10 µs range as well, and its accuracy is ultimately determined by the accuracy of the processor clock.

Figure 7.24 shows a software-generated pulse train. The duration of a full pulse is 25 µs, and the frequency derived from this is $f = 1 / tP = 1 / 25$ µs $= 40$ kHz. Four complete ultrasonic pulses are generated per train. This is to produce a signal that is both strong enough and fast enough, with its 100 µs total duration, to produce accurate spatial resolution.

The complete transmitted and received pulses are illustrated in Figure 7.25. The transmitted pulse arrives at the receiver after a delay of about 25 ms, depending on the distance traversed. Note that the sharply defined transmission packet is returned to the receiver as a soft-edged signal. One reason for this is that, with textured reflective surfaces, different path lengths are traversed by the reflected signals.

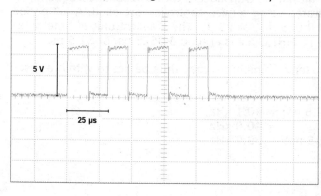

Figure 7.24:
Ultrasonic Transmitter Pulse Train

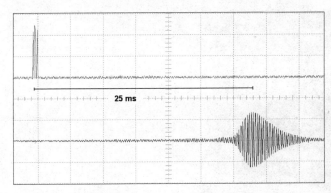

Figure 7.25:
Ultrasonic Transmitted and Received Pulses

The ultrasonic transmitter hardware is shown in Figures 7.26 and 7.27. As the transmission pulse is generated by the Arduino software, the transmitter unit must amplify this 'ping' as much as possible. This is achieved with four transistors operating as a so-called H bridge. Using this, the ultrasonic transmitter generates a signal whose voltage amplitude is twice as high as the power supply voltage. The transmission power is correspondingly roughly quadrupled.

On the receiver side, the signal is processed by two operational amplifiers. In principle, any type of operational amplifier may be used here to provide a sufficient gain-bandwidth product. Very good results were achieved using the Linear Technology LT1007. Since only a unipolar power supply is in use, a virtual ground was formed from two 1 kΩ resistors. After the second amplifier stage, the received signal is rectified, smoothed and finally fed to the Arduino's analog input A0.

Figure 7.26:
Ultrasonic Transmitter

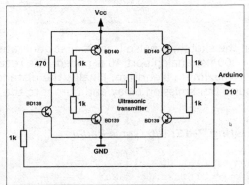

Figure 7.27:
Ultrasonic Transmitter Circuit

7 MEASUREMENT AND SENSORS

Figure 7.28:
Ultrasonic Receiver

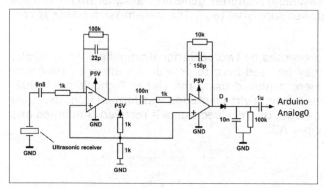

Figure 7.29:
Ultrasonic Receiver Circuit

In the sketch, a pulse train is first generated and output on port pin 10. After a latency of 300 μs, analog port A0 is polled. The received pulse's peak is determined using a simple maximum algorithm. Finally, the distance of the target object is calculated from the pulse transmission delay and output to the serial port.

Listing 7-15: Ultrasonic Radar

```
// Listing 7-15
// Ultrasonic Radar

#include <util/delay.h>
#define F_CPU = 16000000

const float vSound20c = 3432e-7;     // m/us
const unsigned int ave = 30;         // number of averages
const unsigned int pulseLength = 3;  // number of US pulses sent
const float dMin = 0.30;             // minimum measurement distance
```

HOME 'RADAR STATION": DISTANCE MEASUREMENT USING ULTRASOUND 7.16

```
  float sum;                    // total sum of measured values
  float d;                      // result of one distance meas.
  float distance;               // final result for distance
  unsigned int amplitude;       // signal amplitude
  unsigned long tStart;         // start time of measurement
  unsigned long tPeak;          // time of deteceted peak
  unsigned int maxAmplitude;    // maximum of peak
  unsigned long t;              // time counter
  unsigned long tRoundtrip;     // roundtrip time of signal

void setup() {
  Serial.begin(9600);
  pinMode(10, OUTPUT);
}

void loop() {
  for (int n = 0; n < ave; n++) {
    // send ultrasonic pulse train
    // f_US = 40 kHz -> full period = 25 us ->
    // half period = 12.5 us
    for (int i = 0; i < pulseLength; i++) {
      // activate PB2 = Arduino D1
      PORTB = 0b00000100; _delay_us(12.5);
      PORTB = 0b00000000; _delay_us(12.5);
    }

    // restet parameters
    maxAmplitude = 0;
    tStart = micros();
    tPeak = tStart;
    delayMicroseconds(300); // reduce interference

    // scanning for max of received pulse
    for (int i = 0; i < 256; i++) {
      amplitude = analogRead(0);
      t = micros();
      if (amplitude > maxAmplitude) {
        tPeak = t;
        maxAmplitude = amplitude;
      }
    }
    tRoundtrip = tPeak - tStart;
    d = tRoundtrip / 2.0 * vSound20c; // distance is half the size
                                      // of round trip
    sum += d;
  }
  distance = sum / ave; // calculate average
  if (distance > dMin) {
    // send value to interface
    Serial.print("D = "); Serial.print(distance , 2); Serial.println(" m");
```

```
  }
  else {
    Serial.println("Object too close"); // minimum distance reached
  }
  sum = 0;
}
```

8
Timers, Clocks and Interrupts

Since the Arduino is equipped with an integrated crystal, it is well suited to the construction of accurate clocks and timers. With crystal oscillator-based timing, one can achieve accuracy in the parts-per-million range. This means that frequency deviations are not greater than 1:1,000,000. Translated into human terms, this means that the clock will be out by only a few seconds per month.

If you live within 2,000 km of Frankfurt, Germany, even higher precision may be achieved using a DCF77 radio module. This receives time signals from a transmitter near Frankfurt. With it, the maximum time deviation is reduced to less than one second in 10,000 years. That should be more than adequate for everyday needs.

But it's not always just about the maximum achievable precision. The Arduino may also be used to build useful and fun helpers for everyday life. The project in the following section is an example of this.

8.1 Morning and Night Fun: Grand Prix Toothbrush Timer

Brushing one's teeth is an everyday routine task. The toothbrush timer presented in this chapter can add some fun to this chore. Children may find some fun motivation for improved oral hygiene in such a gadget.

For a display, an LED dot matrix is used. We take advantage of the fact that the dot matrix display is capable of rendering not only numbers and letters, but simple graphics as well. The following figures illustrate the symbols used for the cleaning of the left and right upper teeth. Similar symbols are used for the lower teeth. Finally, a smiley indicates the conclusion of the task.

On the hardware side, the dot matrix is connected to the Arduino in the usual manner (see Section 6.4). A piezoelectric transducer is connected between pin 13 and GND. This emits an audible signal when the next mouth quadrant is up.

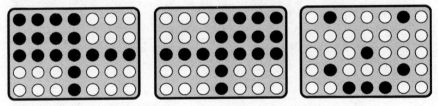

Figure 8.1: *Graphical Displays for Toothbrush Timer*

The cleaning time for each quadrant of the mouth may, of course, be chosen at will, so this is adaptable to individual needs.

To complete the grand prix feel, a red, a yellow and a green LED are also controlled. These exhibit a traffic light sequence:

> red, red & yellow, green, yellow, red

Now, nothing stands in the way of daily oral hygiene.

The entire project may be housed in a small standard enclosure. If a transparent enclosure is used (see Figure 8.2), no recess in the front is necessary, as the display elements are visible through the enclosure. The result is a stylish and splash-resistant device. This latter property should not be underestimated in bathroom applications.

Figure 8.2:
Active Toothbrush Timer

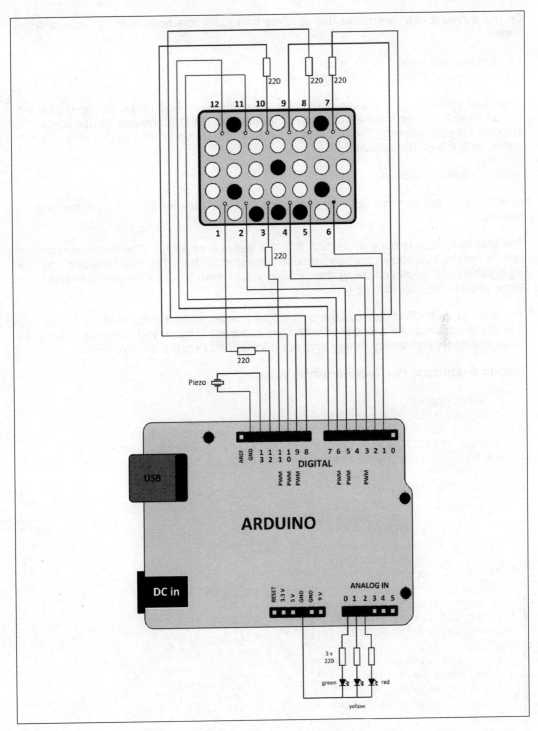

Figure 8.3: Toothbrush Timer Schematic

8 TIMERS, CLOCKS AND INTERRUPTS

On the software side, we make use of some tricks. For example, with the macro definitions

```
#define sbi(PORT, bit) (PORT |= (1 << bit)) // set bit in PORT
#define cbi(PORT, bit) (PORT &= ~(1 << bit)) // clear bit in PORT
```

individual ports may be accessed directly. This approach was taken as the digital pins are all already in use for the dot matrix module, but it's still possible to use the analog inputs as digital outputs. The three traffic light LEDs are controlled in this way. With the above definitions, the following

```
sbi(PORTC, ledRed);
```

directly activates the red LED on processor Port C2 (analog input A2 on the Arduino board).

The graphics, as was done in Section 6.4, are defined as arrays. The numeric representations for the countdown portion are also from that section. The numbers and associated graphics are displayed using the `showMatrix()` and `showQuadrant()` functions. The same applies for `showSmiley()`.

The main program first displays the traffic light phases using a 4-iteration `for` loop. After a beep, the countdown associated with that quadrant is displayed. When all four quadrants are done, the smiley finally appears permanently on the display.

Listing 8-1: *Grand Prix Teeth-brushing Timer*

```
// Listing 8-1
// Grand Prix Teeth-Brushing Timer

#define sbi(PORT, bit) (PORT|=(1<<bit)) // set bit in PORT
#define cbi(PORT, bit) (PORT&=~(1<<bit)) // clear bit in PORT

//      Col 1 2 3 4 5 6 7
// Row
// 1        o o o o o o
// 2        o o o o o o
// 3        o o o o o o
// 4        o o o o o o
// 5        o o o o o o

// define line and column pins
const int col[7] = {7, 6, 5, 4, 3, 2, 1};
const int row[5] = {12, 11, 10, 9, 8};

const byte ledGreen = 0;
const byte ledYellow = 1;
const byte ledRed = 2;
const int maxDelay = 300; // for "delays"

int number; // number on display
```

```
byte quadrant;

// define graphics as 7x5 matrices
const int upLeft[7][5] = {
  {0, 0, 1, 1, 1},
  {0, 0, 1, 1, 1},
  {0, 0, 1, 1, 1},
  {1, 1, 1, 1, 1},
  {0, 0, 1, 0, 0},
  {0, 0, 1, 0, 0},
  {0, 0, 1, 0, 0}
};

const int upRight[7][5] = {
  {0, 0, 1, 0, 0},
  {0, 0, 1, 0, 0},
  {0, 0, 1, 0, 0},
  {1, 1, 1, 1, 1},
  {0, 0, 1, 1, 1},
  {0, 0, 1, 1, 1},
  {0, 0, 1, 1, 1}
};

const int lowLeft[7][5] = {
  {1, 1, 1, 0, 0},
  {1, 1, 1, 0, 0},
  {1, 1, 1, 0, 0},
  {1, 1, 1, 1, 1},
  {0, 0, 1, 0, 0},
  {0, 0, 1, 0, 0},
  {0, 0, 1, 0, 0}
};

const int lowRight [7][5] = {
  {0, 0, 1, 0, 0},
  {0, 0, 1, 0, 0},
  {0, 0, 1, 0, 0},
  {1, 1, 1, 1, 1},
  {1, 1, 1, 0, 0},
  {1, 1, 1, 0, 0},
  {1, 1, 1, 0, 0}
};

const int smiley[7][5] = {
  {0, 0, 0, 0, 0},
  {0, 1, 0, 0, 1},
  {1, 0, 0, 0, 0},
  {1, 0, 1, 0, 0},
  {1, 0, 0, 0, 0},
  {0, 1, 0, 0, 1},
  {0, 0, 0, 0, 0}
```

```
};

const int n[][3] = {
  {31, 17, 31}, // 0
  { 0, 31,  2}, // 1
  {23, 21, 29}, // 2
  {31, 21, 21}, // 3
  {31,  4,  7}, // 4
  {29, 21, 23}, // 5
  {29, 21, 31}, // 6
  {31,  1,  1}, // 7
  {31, 21, 31}, // 8
  {31, 21, 23}  // 9
};

void setup() {
  DDRC = 0b00111111; // Analog Ports as outputs
  for (int i = 0; i < 7; i++) {
    pinMode(col[i], OUTPUT);
  }
  for (int i = 0; i < 5; i++) {
    pinMode(row[i], OUTPUT);
  }
}

void loop() {
  for (int n = 1; n <= 4; n++) {
    sbi(PORTC, ledRed); showQuadrant(n);
    sbi(PORTC, ledYellow); showQuadrant(n);
    cbi(PORTC, ledRed); cbi(PORTC, ledYellow);
    sbi(PORTC, ledGreen);
    tone(13, 440); showQuadrant(n); noTone(13);
    countdown(45, 2); cbi(PORTC, ledGreen); sbi(PORTC, ledYellow);
    countdown(1, 0); cbi(PORTC, ledYellow); cbi(PORTC, ledGreen);
  }
  sbi(PORTC, ledRed);
  while(1) {
    showSmiley();
  }
}

void showMatrix(int displayNumber) {
  // 1s digit
  for (int x = 1; x <= 3; x++) {
    digitalWrite(x, HIGH); PORTB = ~n[displayNumber % 10][x - 1];
    delay(1);
    digitalWrite(x, LOW);
  }
  // 10s digit
  for (int x = 5; x <= 7; x++) {
    digitalWrite(x, HIGH); PORTB = ~n[displayNumber / 10][x - 5];
```

```
      delay(1);
      digitalWrite(x, LOW);
  }
}

void showQuadrant(int quadrant) {
  for (int del = 1; del < maxDelay; del++) {
    for (int x = 0; x < 5; x++) {
      // switch coulmns 0 to 5 for tens digit
      for (int y = 0; y < 7; y++) {
        switch (quadrant) {
          case 1: digitalWrite(col[y], upLeft[y][x]); break;
          case 2: digitalWrite(col[y], upRight[y][x]); break;
          case 3: digitalWrite(col[y], lowLeft[y][x]); break;
          case 4: digitalWrite(col[y], lowRight[y][x]); break;
        }
      }
      // set current column to 0
      digitalWrite(row[x], 0);
      delay(1);
      // reset column
      digitalWrite(row[x], 1);
    }
  }
  for (int y = 0; y < 7; y++) {
    digitalWrite(col[y], 0);
  }
}

void showSmiley() {
  for (int del = 1; del < maxDelay; del++) {
    for (int x = 0; x < 5; x++) {
      // switch coulmns 0 to 5 for tens digit
      for (int y = 0; y < 7; y++)
        digitalWrite(col[y], smiley[y][x]);
      // set current column to 0
      digitalWrite(row[x], 0);
      delay(1);
      // reset column
      digitalWrite(row[x], 1);
    }
  }
}

void countdown(int startNumber, int endNumber) {
  for(number = startNumber; number >= endNumber; number--) {
    for(int n = 1; n < 166; n++) {
      showMatrix(number);
    }
  }
}
```

8 TIMERS, CLOCKS AND INTERRUPTS

8.2 Practical and Accurate: Digital Clock with LED Display

For this project, only a few components are required to build a truly useful, universal tool. In addition to the Arduino, only four resistors, two switches and a 4-digit, 7-segment LED display are needed. From these components, we create a digital clock with individual configuration options.

Of course, digital clocks may be purchased off the shelf. However, DIY offers the joy of creation as well as some more tangible benefits. While commercial clocks offer relatively few design variations, with a DIY clock, one's imagination is the limit. This begins with enclosure selection. One may use a simple and inexpensive plastic enclosure, or handmade creations from fine wood or polished or brushed aluminum may be created. Even unusual materials such as Plexiglas or stainless steel may be used if the right tools are available.

Likewise, the selection of the display brings with it may eye-catching options. While most commercially-available clocks come with red or green LED displays, 7-segment displays are also available in yellow and orange shades, not to mention blue. All in all, a well-designed clock with a very futuristic look may be built.

The customization opportunities don't end with the hardware. By modifying the basic software, almost unlimited customization is possible. For example, an experienced programmer might extend the clock into a multifunctional alarm clock. Display of the date may also be added. Adding the appropriate sensors may enable thermometer or hygrometer functionality. Other examples may be found in the final chapter of this book.

Figure 8.4 shows the basic version of the clock, with four 220 Ω resistors, two switches for setting the time and a 4-digit, 7-segment display.

In the software, two external files are first included. `LED_display.h` provides the usual driver for the display, while `TimerOne.h` is the include file to activate hardware Timer 1. The variables `hrs, mins` and `secs` are then defined. The `tc` variable serves as a seconds counter. The

```
void update_time() { tc++; }
```

function is called once per second by the interrupt, with `tc` representing the clock counter's seconds.

To set the time, the two switches are used – one for the hour and one for minutes.

8.2 PRACTICAL AND ACCURATE: DIGITAL CLOCK WITH LED DISPLAY

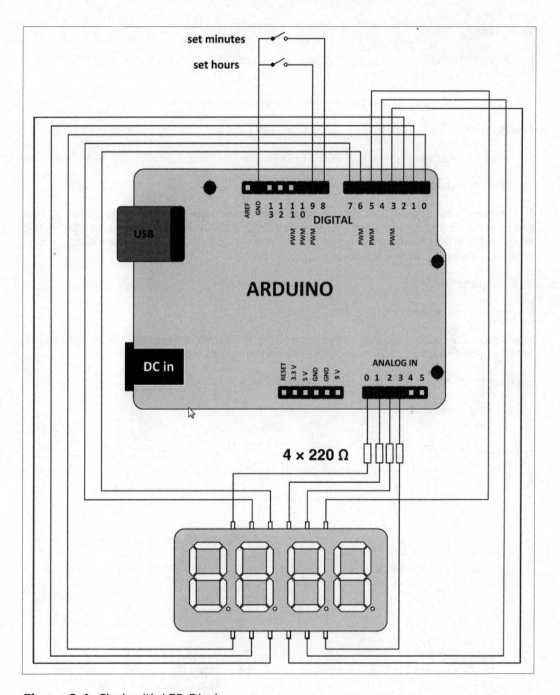

Figure 8.4: *Clock with LED Display*

In `setup()`, the ports necessary for driving the display are set up. The standard port definitions are again used, and the analog inputs are reclassified as outputs.

8 TIMERS, CLOCKS AND INTERRUPTS

Then, with

```
digitalWrite(8, HIGH); // turn on pullup resistor on B0 (Digital 8)
digitalWrite(9, HIGH); // turn on pullup resistor on B1 (Digital 9)
```

the internal pull-up resistors for the switch inputs are enabled.

Finally, with

```
Timer1.initialize(1000000); // Interrupt every 1000000 us = 1.000000 s
```

the timer interrupt is set to exactly one second.

In the main loop, the current time is displayed on the LED display. In the counting routine, the hours and minutes are incremented. When the minute counter exceeds 59, it is reset and the hour counter is incremented. When the hour counter exceeds 23, it, too, is reset, taking the time back to 00:00.

The program line

```
while (secs==tc) refresh();
```

manages a continuous display of the time. Only when the next second interrupt arrives, is the keypad checked, after which the main loop begins again.

Listing 8–2: *Adjustable 7-Segment Display Clock*

```
// Listing 8-2
// Adjustable 7-Segment Display Clock

#include <TimerOne.h>
#include "LedDisplayV3.h" // include display driver

int hr = 10, mn = 0, sec = 0; // set start time
volatile int tc = 0; // timeCounter

#define MIN_KEY_PRESSED (!(PINB & (1 << 0))) // low if Pin B0 is pressed
#define HR_KEY_PRESSED (!(PINB & (1 << 1))) // low if Pin B1 is pressed
void updateTime() {
   tc++;
}

void setup() {
  DDRB = 0b00000000; // port B (pins digital 8 - 13) as input
  DDRC = 0b00001111; // port C 0..3 (pins analog 0 - 3) as output
  DDRD = 0b11111111; // port D (pins digital 0 - 7) as output
  digitalWrite(8, HIGH); // turn on pullup resistor on B0 (Digital 8)
  digitalWrite(9, HIGH); // turn on pullup resistor on B1 (Digital 9)
  Timer1.initialize(1000000); // Interrupt every 1000000 us = 1.000000 s
  Timer1.attachInterrupt(updateTime);
}
```

```
void loop() {
  numberOutput(hr * 100 + mn);
  sec = tc;
  if (60 == sec) {
    tc = 0; sec = 0; mn++;
    if (60 == mn)
    {
      mn = 0; hr++;
      if (24 == hr) {
        hr = 0;
      }
    }
  }

  while (sec == tc) {
    refresh();
  }

  // set minutes
  if (MIN_KEY_PRESSED) {
    if (59 > mn) {
      mn++;
    }
    else {
      mn = 0;
    }
    numberOutput (hr * 100 + mn);
  }

  // set hours
  if (HR_KEY_PRESSED) {
    if (23 > hr) {
      hr++;
    }
    else {
      hr = 0;
    }
    numberOutput (hr * 100 + mn);
  }
}
```

Listing 8-3: *LedDisplayV3.h*

```
// 4x7 segment LED display driver

#include <avr/io.h>
#include <avr/interrupt.h>
#include <util/delay.h>
```

8 TIMERS, CLOCKS AND INTERRUPTS

```c
const int numbers[10][8] = {
  {0, 0, 0, 0, 1, 0, 0, 0}, // 0
  {1, 1, 0, 0, 1, 0, 1, 1}, // 1
  {0, 0, 0, 1, 0, 0, 1, 0}, // 2
  {1, 0, 0, 0, 0, 0, 1, 0}, // 3
  {1, 1, 0, 0, 0, 0, 0, 1}, // 4
  {1, 0, 0, 0, 0, 1, 0, 0}, // 5
  {0, 0, 0, 0, 0, 1, 0, 0}, // 6
  {1, 1, 0, 0, 1, 0, 1, 0}, // 7
  {0, 0, 0, 0, 0, 0, 0, 0}, // 8
  {1, 0, 0, 0, 0, 0, 0, 0}  // 9
};

volatile int D1, D2, D3, D4;
volatile uint8_t activeDigit = 0;

#define sbi(PORT, bit) (PORT |= (1 << bit))
#define cbi(PORT, bit) (PORT &= ~(1 << bit))

void digit(int value, int pin) {
  sbi(PORTC, pin);
  // cbi(PORTC, pin);    // enable for common cathode
  PORTD = 0b11111111;
  // PORTD = 0b00000000; // enable for common cathode
  for (int i = 0; i <= 7; i++) {
    if (0 == numbers[value][i]) {
      cbi(PORTD, (i));
      // sbi(PORTD, (i)); // enable for common cathode
      _delay_ms(1);
      sbi(PORTD, (i));
      // cbi(PORTD, (i)); // enable for common cathode
    }
  }
}

void numberOutput (uint16_t number) {
  D1 = number / 1000; number %= 1000;
  D2 = number / 100;  number %= 100;
  D3 = number / 10;   number %= 10;
  D4 = number;
}

void refresh() {
  PORTC = 0b00000000; // Digits off
  // PORTC = 0b00001111; // enable for common cathode
  if (0 == activeDigit) {
    digit(D1, PORTC0);
  }
  if (1 == activeDigit) {
    digit(D2, PORTC1);
  }
```

```
    if (2 == activeDigit) {
      digit(D3, PORTC2);
    }
    if (3 == activeDigit) {
      digit(D4, PORTC3);
    }
    activeDigit ++;
    if (4 == activeDigit) {
      activeDigit = 0;
    }
  }
}
```

8.3 Who's Faster? A Reaction Timer

A person's reaction time plays a crucial role in everyday situations. In road traffic, for example, reaction time may be a matter of life and death. Typical reaction timing differs very slightly from person to person. Much larger differences are due to the particular psychological state of an individual. Alcohol and drugs have a clear effect here. However, lack of sleep or other stressful situations also have an adverse effect on a person's reaction timing.

The reaction timer presented here provides a rough indication of the speed at which one reacts to a particular event. The event, in this case, is an LED display turning from red to green. When the red LED goes out and the green one lights up, one attempts to press a button as quickly as possible. The time between the event and the key press is measured and shown on the LED display. The number displayed indicates the elapsed time in thousandths of a second, so a display of '0327' indicates 0.327 seconds.

At each attempt, the signal changes after a different amount of time. This is so that players can't improve their apparent reaction time by becoming accustomed to a constant time before the light changes. The reaction timer's circuit diagram is shown in Figure 8.5.

8 TIMERS, CLOCKS AND INTERRUPTS

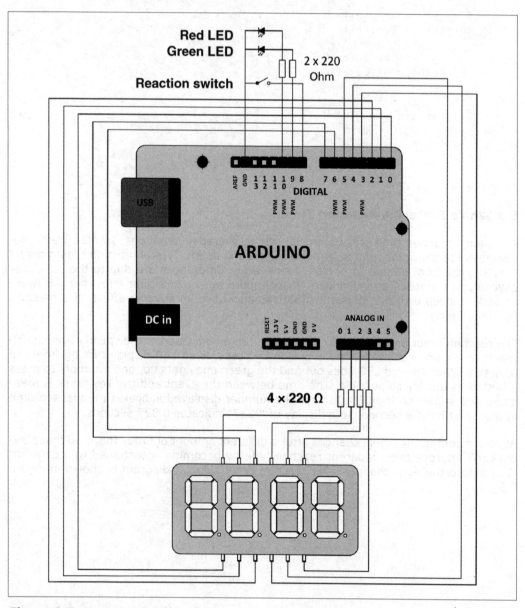

Figure 8.5: Reaction Timer

In the program, after including the `LED_display.h` and `TimerOne.h` files, the pins for the two LEDs and the switch are set. The port definitions for the LED display are already familiar from the previous section.

In the main loop, first the red LED is activated. With the help of the `random()` function, a random delay time is defined. The associated delay loop then follows. If the key is pressed before the signal change, a disqualification is indicated by '9999'.

After the delay, the red LED is deactivated, and the green activated. In a further loop, the time between green LED activation and the key press is measured. This time corresponds to the reaction time. It is displayed on the LED display in the above-mentioned format.

Listing 8-4: Reaction Timer

```
// Listing 8-4
// Reaction Timer

#include <TimerOne.h>            // include timer
#include "LedDisplayV3.h"        // include display driver

const byte button = 8;
const byte ledRed = 10;
const byte ledGreen = 9 ;
int tc; // time counter
int randomDelay;

#define KEY_PRESSED (!(PINB & (1 << 0))) // low if button on Pin B0 is pressed

void setup() {
  pinMode(ledRed, OUTPUT);
  pinMode(ledGreen, OUTPUT);
  pinMode(button, INPUT);
  digitalWrite(button, HIGH); // turn on pullup on B0 (Digital 8)
  DDRC = 0b00001111; // port C 0-3 (pins analog 0-3) as output
  DDRD = 0b11111111; // port D (pins digital 0-7) as output
  randomSeed(100);
}
void loop() {
  digitalWrite(ledRed, HIGH);
  randomDelay = random(10, 100);
  for (int i = 0; i < randomDelay; i++) {
    if (KEY_PRESSED) {
      numberOutput (9999);
      for (int i=0; i < 3000; i++) {
        refresh();
      }
    }
    delay(100);
  }
  digitalWrite(ledRed, LOW);
  digitalWrite(ledGreen, HIGH);
  while (!KEY_PRESSED) {
    tc++; delayMicroseconds(1000);
  }
  digitalWrite(ledGreen, LOW);
  if (tc > 9999) {
```

```
      tc = 9999;
  }
  numberOutput(tc);
  for (int i = 0; i < 3000; i++) {
    refresh();
  }
  tc = 0;
}
```

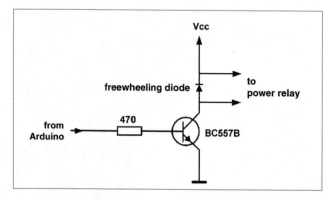

Figure 8.7:
Transistor Driver Stage

8.4 'Timerino': Universal Timer with a 7-Segment Display

Timers allow specific, scheduled tasks to be run. With the timer presented here, times of up to 99 minutes and 59 seconds, or almost one hour and 40 minutes, may be set. The selected time period, as well as the time remaining, may be read from the 4-digit, 7-segment display.

Once the time has elapsed, the display blinks 00:00. In addition, digital output 13 is switched. Thus, after the selected duration has elapsed, a relay may be triggered to control any electrical appliance. How one may use suitable relays to operate line voltage devices is described in the next section.

The applications of such a timer are varied. For example, a television may be turned off automatically. This is very useful for people who tend to fall asleep in front of the box. Other examples are tanning machines, dark room equipment and PCB production exposure devices. A timer may also be used to control battery charging times. It would serve as an additional safeguard against overcharging, in addition to the usual monitoring of charge current and charged voltage.

In the program, the functions of the controls are first defined. In `setup()`, the pins required for LED display control are configured. Then, Timer 1 is initialized with an interrupt interval of 1,000,000 us = 1,000 ms = 1 s.

In the main program, the start time may be adjusted using the buttons for minutes and seconds. After pressing the 'Start' button, the time starts counting down, and the digital display is updated each second.

Extensions and exercises:

- Modify the program so that a long-term timer is created that can manage times of up to 99 hours and 59 minutes, i.e. almost 4 days and 4 hours, at a resolution of one minute.

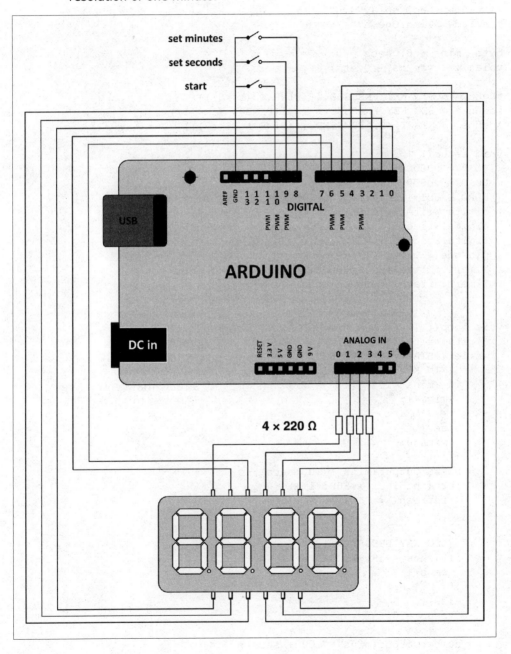

Figure 8.6: Universal Timer

Listing 8-5: *Timer with 4x7-Segment Display*

```
// Listing 8-5
// Timer with 4x7-Segment Display

#include "LedDisplayV4.h"
#include "TimerOne.h"

byte   mins = 0, secs = 0;
volatile byte ts = 0, tm;

#define    MIN_KEY_PRESSED (!(PINB & (1<<0)))
#define    SEC_KEY_PRESSED     (!(PINB & (1<<1)))
#define    START_KEY_PRESSED (!(PINB & (1<<2)))

void setup() {
  DDRB = 000000;
  DDRC = 0b00001111;
  DDRD = 0b11111111;
  digitalWrite(8, HIGH);
  digitalWrite(9, HIGH);
  digitalWrite(10, HIGH);
  pinMode(13, OUTPUT);
  Timer1.initialize(1000000);
  Timer1.attachInterrupt(updateTime);
}

void loop() {
  digitalWrite(13, LOW);
  while (!START_KEY_PRESSED) {
    if (MIN_KEY_PRESSED) {
      if (mins < 59) {
        mins++;
      }
      else {
        mins=0;
      }
      numberOutput(mins * 100+secs);
      for (int i=0; i<300; i++) {
        refresh();
      }
    }
    if (SEC_KEY_PRESSED) {
      if (secs < 59) {
        secs++;
      }
      else {
        secs = 0;
      }
      numberOutput(mins * 100 + secs);
      for (int i = 0; i < 300; i++) {
```

```
      refresh();
    }
  }
  numberOutput (mins*100+secs);
  refresh();
}
ts = secs;
tm = mins;
while ((tm > 0)| (ts > 0)) {
  numberOutput(tm * 100 + ts);
  refresh();
}
digitalWrite(13, HIGH);
numberOutput(0000);
for (int i = 0; i < 3000; i++) {
  refresh();
}

while (!START_KEY_PRESSED) {
  for (int i = 0; i < 300; i++) {
    refresh();
  }
  numberOutput (0000);
  delay(300);
}
while (START_KEY_PRESSED);
}

void updateTime() {
  if (ts > 0) {
    ts--;
  }
  else {
    ts=59;
    if (tm > 0) {
      tm--;
    }
  }
}
```

8.5 Plug-in Timer to Make Life Easier

When timed switching of electrical appliances is required, a power relay is necessary. These devices are available from various electronics mail order companies, such as ELV (see Supplier Directory). One should avoid trying to create DIY high-voltage switching devices, as there is a risk factor associated with the lethal voltages.

There are commercial devices that may be driven directly by 5 volts. On the high power side, they have a safety plug and a grounded outlet. The high- and low-voltage circuits are fully separated and all of the relevant safety standards are adhered to. Thus, in the proper application of these devices, danger to life and limb is avoided.

8 TIMERS, CLOCKS AND INTERRUPTS

Depending on the design of the power control device, it may be controlled directly from a processor pin. This is the case when 5 volts at a maximum current of 20 mA are sufficient to switch the device. For devices with built-in electromechanical relays, higher switching voltages or higher currents may be required. Should this be the case, an additional transistor driver stage may be necessary.

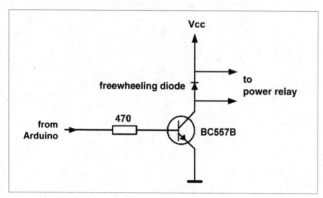

Figure 8.7:
Transistor Driver Stage

By using a suitable power switching device, nothing stands in the way of all of the practical applications mentioned in the previous section.

8.6 Atomic Precision: The DCF77 Radio Clock

A long wave DCF77 transmitter exists near Frankfurt, Germany. It is used to supply radio-controlled clocks in Germany with the exact current time. Central European Time (or Central European Summer Time) is transmitted on its one-second beats. The time base is generated by a special clock at the transmitter site, which is in turn based on the atomic clocks at the PTB (Physikalisch-Technische Bundesanstalt) in Braunschweig. Depending on the time of day and time of year, the signal may be received at distances of up to 2,000 km.

The complete coded time signal is transmitted once every minute. The coding is carried out over the pulses transmitted each second. The signal is based on the following bit definitions:

- A pulse length of 100 ms or 0.1 s = logical '0'.

- A pulse length of 200 ms or 0.2 s = logical '1'.

For synchronization purposes, a 59th bit is not transmitted. The bits in a complete, one-minute pulse sequence have the following meanings:

ATOMIC PRECISION: THE DCF77 RADIO CLOCK 8.6

Second	Purpose
0	Indicates beginning of minute. Always 0.
1 – 14	Encrypted weather data.
15	Abnormal transmitter operation indicator.
16	Summer time announcement.
17	Active during CEST (Summer).
18	Active during CET (Winter).
19	Leap second announcement.
20	Start of time indicator. Always 1.
21 – 27	Minute bits 1, 2, 4, 8, 10, 20 and 40.
28	Even parity check bit for minutes.
29 – 34	Hour bits 1, 2, 4, 8, 10 and 20.
35	Even parity check bit for hours.
36 – 41	Day of month bits 1, 2, 4, 8, 10 and 20.
42 – 44	Day of week bits 1, 2 and 4.
45 – 49	Month bits 1, 2, 4, 8 and 10.
50 – 57	2-digit year bits 1, 2, 4, 8, 10, 20, 40 and 80.
58	Even parity check bit for date.
59	Minute mark. No data sent.

Table 13: *Structure of the DCF77 Signal*

Figure 8.8 shows a portion of the DCF77 signal, as measured at the output of a standard receiver.

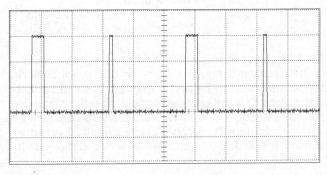

Figure 8.8:
Part of the DCF77 Signal

For an initial test of the receiver, the pulse train may be fed to an LED. Since most of these modules can only drive very small currents, an additional transistor stage is required. Figures 8.9 and 8.10 show a suitable circuit diagram and test rig, respectively.

8 TIMERS, CLOCKS AND INTERRUPTS

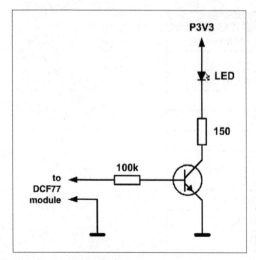

Figure 8.9:
Circuit Diagram for DCF77 Module with LED and Transistor Stage

Figure 8.10:
DSF77 Test Rig with LED and Transistor Stage

Using this test rig, the LED should blink once every second. The different pulse lengths of either 0.1 or 0.2 seconds should also be clearly distinguishable. If the LED is not blinking, or blinks irregularly, the ferrite antenna may need to be better aligned. It should also be noted that the DCF receiver is very sensitive to interference from laptop power supplies, computers and other sources of spurious radiation. It may be necessary to separate the receiver by up to a meter from other equipment to receive clean signals. Even moving the antenna a certain distance from the Arduino itself may improve signal quality.

Note

Some DCF77 modules may only be operated with a 3.3 V supply. A module of this type should be powered by the corresponding Arduino pin. More detailed information should be in the module's datasheet.

8.7 Output of Time and Date to the Serial Interface

In our first version of the DCF77 clock, the time will be output to the serial interface. The DCF77 module may be connected directly to the Arduino. Most of these modules have four pins:

1) VCC (NOTE: often only 3.3V instead of 5 V)
2) GND
3) Signal output
4) Standby

Depending on the module, VCC is connected either to the Arduino's 5V or to its 3.3V output. In addition, of course, the grounds should be connected. The module's signal output should be connected to the Arduino's A1 analog input.

Many DCF77 boards also have a fourth pin. This allows the module to be placed in either active or standby mode. To do so, this pin must be connected to either GND or 5V. Find this detail in the datasheet. Alternately, just try it out.

On the software side, another helpful library file is available. This may be downloaded from

lietaer.be.dotnet39.hostbasket.com/arduino/Dcf77.zip

In the sketch, the file is included in the usual manner, with `#include "Dcf77.h"`. In `setup()`, only the serial interface need be initialized.

In the main loop, the current time is retrieved using the `Dcf77.getDateTime()` function. Should a valid string be returned, this is output to the serial interface. DCF77 data has the following format:

```
DCF77DATA format 2010120662110CET
CCYYMMDDdhhmmTTTT
```

    `CCYY`: four-digit year, e.g. 2010
    `MM`: two-digit month, e.g. 12 for December
    `DD`: Two-digit day of the month, e.g. 06
    `d`: single-digit day-of-week code, e.g. 1 for Monday
    `HH`: two-digit hour, e.g. 21 for 9 PM
    `mm`: two-digit minute, e.g. 10
    `TTTT`: up-to-four-character supplemental information, e.g. CET for Central European Time

The decoded data is output directly to the serial interface, and may be monitored using Serial Monitor or another terminal program.

Finally, the LED on pin 13 is switched in time with the DCF77 signal. This is very useful for evaluating the reception quality of the circuit. The blinking LED also helps when it comes to aligning the antenna and finding a suitable placement for reception.

8 TIMERS, CLOCKS AND INTERRUPTS

Listing 8-6: *DCF77 Decoder with Serial Output*

```
// Listing 8-6
// DCF77 Decoder with Serial Output

// DCF77DATA format 2010120662110CET
// CCYYMMDDdhhmmTTTT

#include "Dcf77.h"

Dcf77 Dcf77(1);  // use analog input pin 1 on Arduino

int ledPin = 13;       // LED connected to digital pin 13

void setup() {
  Serial.begin(9600);
  pinMode(ledPin, OUTPUT);
}

void loop() {
  const char *v = Dcf77.getDateTime();
  if (0 != strcmp(v, "DCF77POLL")) {
    Serial.println(v);
    Serial.print("Year: "); Serial.print(v[10]); Serial.print(v[11]);
    Serial.print(v[12]);Serial.println(v[13]);
    Serial.print("Date: "); Serial.print(v[14]); Serial.print(v[15]);
    Serial.print("."); Serial.print(v[16]); Serial.println(v[17]);
    Serial.print("Day: "); Serial.println(v[18]);
    Serial.print("Time: "); Serial.print(v[19]); Serial.print(v[20]);
    Serial.print(":"); Serial.print(v[21]);Serial.println(v[22]);
    Serial.print("Info: "); Serial.print(v[23]); Serial.print(v[24]);
    Serial.print(v[25]); Serial.println(v[26]);
    Serial.println();
  }
  if (100 < analogRead(1)) {
    digitalWrite(ledPin, HIGH);
  }
  else {
    digitalWrite(ledPin, LOW);
  }
}
```

8.8 Stand-Alone DCF77 Clock with LCD Display

Should one not wish to turn on a PC just to see the time, the time data may be sent to an LCD instead of the serial interface. In this way, we obtain a complete, self-contained and highly accurate clock. Again, the advantage over a commercially available clock is the opportunity for endless customization. For example, an alarm may be integrated. One is limited not only to the time of day here, but we can make use of the day of the week to determine alarm times. Aside from simple alarm sounds, one may attach a radio module or MP3 player, etc. More ideas may be found in *Extensions and exercises* at the end of this section.

Listing 8-7: *DCF77 Clock with LCD Display*

```
// Listing 8-8
// DCF77 Decoder with LCD Display

// DCF77DATA format 2010120662110CET
// CCYYMMDDdhhmmTTTT

#include <LiquidCrystal.h>
#include "Dcf77.h"
// create LCD object, specifying pins
LiquidCrystal lcd(12, 11, 5, 4, 3, 2);

Dcf77 Dcf77(1); // use analog input 1 on Arduino
int ledPin = 13; // LED connected to digital pin 13

void setup() {
  Serial.begin(9600);
  pinMode(ledPin, OUTPUT);
  Serial.println("Starting...");
  lcd.begin(16, 4); // initialize LCD to 16 columns, 4 rows
  delay(1000);      // delay for slower LCD modules
  lcd.setCursor(3, 0);
  lcd.print("DCF77 Clock");
  delay(1000);
  lcd.clear();
}

void loop() {
  const char *v = Dcf77.getDateTime();
  if (0 != strcmp(v, "DCF77POLL")) {
    Serial.println(v);

    lcd.setCursor(3, 0);
    if ('1' == v[18]) lcd.print ("Monday ");
    if ('2' == v[18]) lcd.print ("Tuesday ");
    if ('3' == v[18]) lcd.print ("Wednesday ");
    if ('4' == v[18]) lcd.print ("Thursday ");
    if ('5' == v[18]) lcd.print ("Friday ");
    if ('6' == v[18]) lcd.print ("Saturday ");
    if ('7' == v[18]) lcd.print ("Sunday ");

    lcd.setCursor(3, 1);
    lcd.print(v[16]); lcd.print(v[17]); lcd.print(".");
    lcd.print(v[14]); lcd.print(v[15]); lcd.print(".");
    lcd.print(v[10]); lcd.print(v[11]); lcd.print(v[12]); lcd.
print(v[13]);
```

8 TIMERS, CLOCKS AND INTERRUPTS

```
    lcd.setCursor(19, 0);
    lcd.print(v[19]); lcd.print(v[20]); lcd.print(":");
    lcd.print(v[21]); lcd.print(v[22]);

    lcd.setCursor(19, 1);
    lcd.print(v[23]); lcd.print(v[24]); lcd.print(v[25]); lcd.println(v[26]);
  }
  if (100 < analogRead(1)) {
    digitalWrite(ledPin, HIGH);
  }
  else {
    digitalWrite(ledPin, LOW);
  }
}
```

Extensions and exercises

- Program a universal alarm clock that goes off at 07:00 on weekdays, for example, but only at 09:00 on weekends.

- Experiment with it to wake you to:
    simple tones
    short melodies

- Connect a radio module to a port and create your own clock radio.

- Using an MP3 player module or MP3 shield, one may awake to one's favorite song every day.

- With a 230 V relay, lamps and other electrical appliances may be controlled.

9
Interfaces

In digital devices, data must often be exchanged between different modules. For this purpose, various bus systems have been developed. Each of these has specific advantages and disadvantages. Some buses are particularly immune to interference, while others can traverse long distances. In the microcontroller field, one bus is of special note: the I²C bus ("I squared C").

9.1 Universal and Simple: The I²C Interface

What's special about the I²C interface is that it gets by with minimal hardware complexity. I²C stands for inter-IC, i.e. inter-integrated circuit. It was specified and developed by Philips Semiconductors (now NXP Semiconductors). For this reason, it is found in many consumer electronics devices.

The I²C bus offers a synchronous, serial, two-wire interface between a master and one or more slaves. On one wire, SCL (Serial CLock), the clock is transmitted, while on the other, SDA (Serial DAta), the data is exchanged. Data may be sent from master to slave or vice-versa, but the clock is always generated by the master.

In principle, it is possible to have more than one bus master. We then have a so-called multi-master system. In such a case, precaution must be taken against collisions, i.e. a master may only communicate on the bus when the bus is not already engaged by another master. This mode is relatively rare, so it won't be described in any detail here.

The I²C bus has four speeds:

- 100 kHz clock: Standard Mode
- 400 kHz clock: Fast Mode
- 1.0 MHz clock: Fast Mode Plus
- 3.4 MHz clock: High-Speed Mode

Usually, clock rates of up to 1 MHz are supported. The basic structure of the I²C bus system is shown in Figure 9.1.

9 INTERFACES

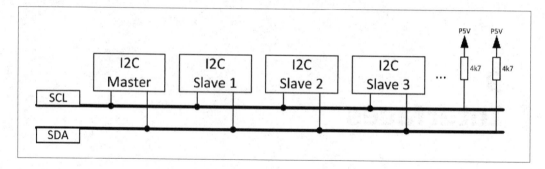

Figure 9.1: *Basic Structure of the I²C Bus*

Since all of the ICs are connected in parallel to the same bus, an addressing scheme is necessary to select an individual receiver. All I²C devices may be selected via a 7-bit address. There is a newer, 10-bit addressing scheme available, which may be used alongside the 7-bit system. While the 7-bit scheme allows for up to 128 addresses, a 10-bit address may represent 1024 possible addresses. Since the 10-bit version is rarely used in the non-industrial environment, we won't look at that here, either.

The following diagram shows an I²C message bit sequence. The 8th bit (R/W) at the end of the address byte indicates whether a read or a write is required to take place. The bytes that come after the address byte are dependent on the device.

	Address			Device Subaddress			R/W		Data							
Start	0	1	0	0	x	x	x	1	ACK	x	x	x	x	x	x	x	x	Stop

Figure 9.2: *I²C Message Bit Sequence*

The I²C bus's SCL clock signal is generated by the microcontroller. Obviously, not every controller's clock will be exactly 100 kHz or 400 kHz, but this is not critical, as the I²C clock is not highly frequency sensitive. What's important is that the actual clock frequency is not higher than the maximum frequency that the connected devices can handle.

If data is to be transmitted over the bus, the controller must first enable this. For anti-collision purposes, the master checks the voltage levels on the two lines, SCL and SDA. Only when both lines are at the HIGH level will the master begin a communication. On the I²C bus, pull-up resistors are required in order to keep both bus lines HIGH when they are not in use. These are connected externally from 5V to the bus. The typical value for such resistors is 4.7 kΩ. Because the ATmega328P has internal pull-ups, no external resistors are required for this application. However, there is no harm in adding external pull-ups to improve signal integrity.

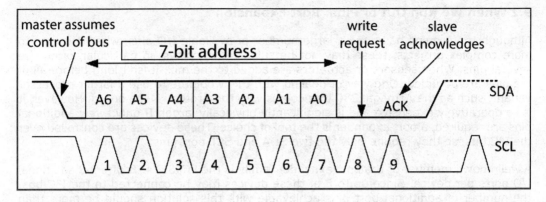

Figure 9.3: I²C Timing

To start communication, the master pulls SDA to ground, while SCL is still HIGH. The master is then in control of the bus, and all connected I²C devices recognize this as a START signal.

Data transmission may now begin with addressing. This determines which slave will be selected for communication with the master. Each I²C-compatible device or IC has a fixed address. In addition to this fixed address assigned to each device, most I²C-compatible components have a portion of the address set aside to be freely selectable. Using some external pins, identical devices may be set to different addresses. For example, should three pins be available for custom addressing, the device may have one of 8 different addresses, depending on which individual pins are connected to GND and which to 5V. Should a device have a fixed internal address of `101 0xxx`, then, by switching address pins A0, A1 and A2, an address in the range `101 0000` – `101 0111` is achievable, i.e. if all three pins are tied to GND, the chip will have an address of `101 0000` – the first of 8 possible addresses, allowing up to 8 identical devices to be addressable on the same bus.

For data transfer from master to slave, the master sends out a 7-bit long slave address, after which it releases the SDA line and waits for an acknowledgement. All of the slaves on the bus compare this with their own address. If the address matches that of a specific device, that slave pulls SDA LOW, signaling that it's ready for further communication. Confident in the knowledge that there is a device with that address present, the master then begins sending 8-bit long data words over the bus, which are again each acknowledged by the slave with an ACK.

If data is to be read from the slave, the master sends the 7-bit address again, although the 8th bit will contain a 1, which indicates a read operation. Again, the addressed slave acknowledges with an ACK. The master sends out 8 clock pulses, to which the slave returns 8 data bits in synchronization. The master reads each of these bits, and then responds with its own ACK.

At the end of the transfer, the bus must be released. To signal the end of the exchange, the master releases the SDA line from LOW to HIGH, while the SCL line is still in the HIGH state. This is known as a STOP condition. The bus is then free and available for further communication.

9.2 When We Run Out of Pins: Port Expansion

Although the Arduino has quite a large number of available I/O pins, when it comes to more complex projects, these may not be sufficient. Displays, in particular, often use several pins. When sensors or actuators are added to the mix, it isn't long before all of the pins are occupied. One way of dealing with this would be to use a larger processor variant, such as the ATmega1280 that comes with the Arduino Mega board. However, it is comparatively expensive and is significantly physically larger. If only a few additional pins are required, a port expander is the tool of choice. These devices are controlled over the I²C bus, so they require only the two SDA and SCL port pins.

A well known member of the port expander family is the PCF8574. It offers 8 additional I/O ports per device. Since up to 8 of these devices may be connected to the I²C bus, the number of additional port pins achievable with this solution should be more than adequate for almost any purpose.

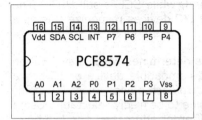

Figure 9.4:
PCF8574 Port Expander

9.3 Mega 24-LED Chaser Using the PCF8574

To introduce the concept of port expanders on the I²C bus, a mega LED chaser will be demonstrated here. With three PCF8574s, we will control 24 LEDs. The following figures show the circuit diagram and construction photograph.

Important note
The LEDs used here are the 5 V models we've mentioned before. For ordinary LEDs, the usual series resistors will also be required.

MEGA 24-LED CHASER USING THE PCF8574   9.3

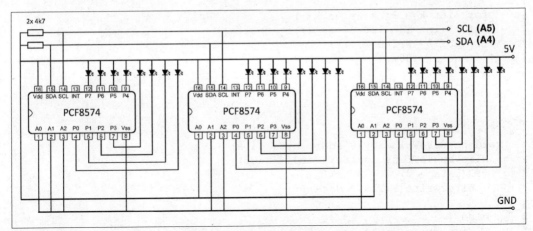

Figure 9.5: Schematic of a Mega Chaser with Three PCF8574 Port Expanders

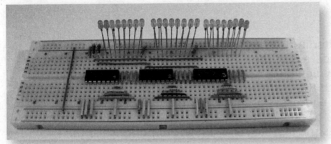

Figure 9.6: Construction of the Mega Chaser

Listing 9-1: PCF8574 Mega LED Chaser

```
// Listing 0901
// PCF8574 Mega LED Chaser

#include <Wire.h>
#define addr0 0b00100000 // I2C address of 1st PCF8574
 // addr0 + 1 = I2C address of 2nd PCF8574
 // addr0 + 2 = I2C address of 3rd PCF8574

void setup() {
  Wire.begin();
}

void loop() {
  for (int k = 0; k <= 2; k++) {
    for (int i = 1; i <= 128; i <<= 1) {
      pcf8574Write(k, i);
      delay(100);
    }
    pcf8574Write(k, 0);
  }
```

9 INTERFACES

```
      for (int k = 2; k >= 0; k--) {
        for (int i = 128; i >= 1; i >>= 1) {
          pcf8574Write(k, i);
          delay(100);
        }
        pcf8574Write(k, 0);
      }
    }

    void pcf8574Write(byte addr, byte data) {
      Wire.beginTransmission(addr0 + addr);
      if (data > 0) {
        Wire.write(~(128 / data));
      }
      else {
        Wire.write(255);
      }
      Wire.endTransmission();
    }
```

In the sketch, the I²C library is first included. This is done with the `#include <Wire.h>` directive. Subsequently, the address of the first module is `#define`d. Since the first PCF8574 has all three of its freely definable address bits tied to ground, its address will be binary `0010 0000`, of which the first five bits are fixed internally in the chip. The last three bits are `000`. Since the addresses are incremented for each of the other two ICs, their last three bits will be `001` and `010`, respectively. The main function consists of two loops. The first loop causes the light to run from left to right, while the second returns it.

In the loops, the individual I²C devices are addressed in sequence. Since the addresses follow each other, there is no need to explicitly `#define` them — they are addressed using a simple increment. In the second loop, the loop index, `i`, is doubled with each pass. Thus, a sequence of powers of two is generated:

                1, 2, 4, 8, 16, 32, 64, 128

or, in binary,

```
    0000 0001
    0000 0010
    0000 0100
    0000 1000
    0001 0000
    0010 0000
    0100 0000
    1000 0000
```

In our setup, this corresponds with a left-to-right running light. This sequence is passed to

```
    void PCF8574_Write(byte addr, byte data)
```

There, the value 128 is divided by each of the numbers in the sequence and inverted:

```
Wire.write(~(128 / data));
```

This makes the light to move right with each higher number, corresponding with the notion that numbers get larger toward the right side of a number line. Of course, this behavior is achievable in many other ways.

Since all blocks are accessed sequentially in the outer loop, the light spot eventually runs through all 24 LEDs.

In the second loop, the numbers are counted down, that is, the given number is halved each time. Thus, the spot moves from right to left.

9.4 Hexadecimal Debugger Display Using a 2-Digit, 7-Segment Display

As we've seen with the 4-digit, 7-segment display, it uses up quite a bit of resources in multiplexed operation. Not just the controller's 14 I/O ports are required, but also either an internal timer or a significant amount of program time for the refreshes is required. These resources are then unavailable to other applications.

An LED display is more economical with processor resources when it's driven via the I²C bus. Then, only the two I²C lines, SDA and SCL, are required. If the I²C bus is already in use, the display can piggy back on these lines, and no further pins are required.

The eight I/O pins on the PCF8574 are just right for a 7-segment display (with decimal point). Since no multiplexing takes place, one simply has to send the desired number once, via the I²C bus, to the display unit. Continuous refreshing is not required, and thus neither the timer nor interrupts are required either.

With a two-digit common-anode display and two I²C port expander chips, a very practical hexadecimal display may be constructed for debugging purposes. This circuit is very useful when it comes to debugging microcontroller software. When this small device is connected to the microcontroller via the I²C interface, one may very easily display the content of registers or port statuses from within a program. In this way, many bugs may be tracked down very quickly. Another advantage is that virtually no additional processor resources are required.

The display of register content would be in hexadecimal format, i.e. each single display digit will represent a value from 0 to 15. The values greater than 9 are represented by letters of the alphabet.

Decimal	0	1	2	3	4	5	6	7	8	9	10	11	12	13	14	15
Hexa-decimal	0	1	2	3	4	5	6	7	8	9	A	b	C	d	E	F

Table 14: Decimal and Hexadecimal Numbers

Here, one takes advantage of the fact that the letters A to F are quite legibly presentable on a 7-segment display. While one digit may display up to 16 different values, two of

9 INTERFACES

them will display up to 16 × 16, or 256 values (0 to 255_{dec}, or 00_{hex} - FF_{hex}).

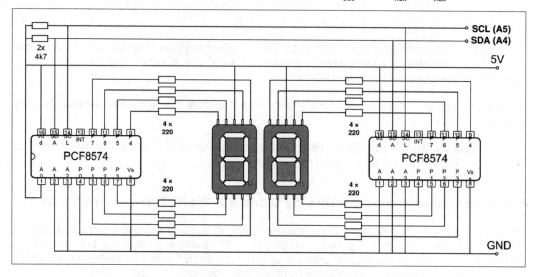

Figure 9.7: Schematic for Hexadecimal Debugger Display

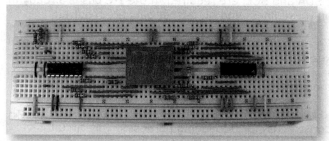

Figure 9.8:
Hex Debugger Display
Construction Example

Listing 9-2: PCF8574 Hexadecimal Debugger Display

```
// Listing 9-2
// PCF8574 Hexadecimal Debugger Display

#include <Wire.h>
#define addr0 0b00100000   // I2C address of PCF8574A - A0, A1 and A2
== 0
#define addr1 0b00100001   // I2C address of PCF8574A - A0 == 1, A1 and
A2 == 0

const byte digit[] = {
  //bagfedcX       bagfedcX       bagfedcX       bagfedcX
  0b00010001, 0b01111101, 0b00100011, 0b00101001, // 0 1 2 3
  0b01001101, 0b10001001, 0b10000001, 0b00111101, // 4 5 6 7
  0b00000001, 0b00001001, 0b00000101, 0b11000001, // 8 9 A b
```

HEXADECIMAL DEBUGGER DISPLAY USING A 2-DIGIT, 7-SEGMENT DISPLAY 9.4

```
    0b10010011, 0b01100001, 0b10000011, 0b10000111, // C d E F
    0b11101111 // -
};

void pcf8574Write(byte addr, byte data) {
  Wire.beginTransmission(addr);
  Wire.write(data);
  Wire.endTransmission();
}

void writeDec (byte x) {
  // if it's between 0 and 99, display it
  if (x < 100) {
    pcf8574Write(addr1, digit[x / 10]);
    pcf8574Write(addr0, digit[x % 10]);
  }
  // else just display --
  else {
    pcf8574Write(addr1, digit[16]);
    pcf8574Write(addr0, digit[16]);
  }
}

void writeHex (byte x) {
  pcf8574Write(addr1, digit[x / 16]);
  pcf8574Write(addr0, digit[x % 16]);
}

void setup() {
  Wire.begin();
}

void loop() {
  for (int i = 0; i < 256; i++) {
    writeHex(i); delay(500);
    // writeDec(i); delay(500);
  }
}
```

In this sketch, after including the Wire library and setting the I²C addresses, the number codes for the 7-segment display are initialized. The procedure corresponds to the principles laid out in Section 6.2. In addition to the numbers 0 to 9, the letters A to F are also defined.

The function

    void PCF8574_Write(byte addr, byte data)

transfers the digit data to the device referenced by the address byte. With the two functions

```
void writeDec(byte x)
```

and

```
void writeHex(byte x)
```

a simple decimal or full hexadecimal representation may be selected by uncommenting the relevant code. In the main loop, counting is done, either from 00_{hex} to FF_{hex} or from 0_{dec} to 99_{dec}.

9.5 LCD Control via I²C Using the PCF8574

LC displays may also be connected to the PCF8574. Since these are widely used, the opportunity to save on processor pins will be described here. Another advantage of this method is that several displays may be connected to a controller with ease. In this case, each display is connected to its own PCF8574. When the individual port expanders are assigned their own addresses, up to 8 LCDs may be connected to one Arduino. The individual LCD displays are then individually addressable by their respective I²C addresses. The advantage of I²C communication could not be clearer here: all 8 LCDs are driven by only two I/O pins.

Here, again, a suitable library is available for download:

www.xs4all.nl/~hmario/arduino/LiquidCrystal_I2C/LiquidCrystal_I2C.zip

Our sketch is very compact, thanks to this library. After including the libraries using

```
#include <Wire.h>
#include <LiquidCrystal_I2C.h>
```

one need only set the I²C address and display format using

```
LiquidCrystal_I2C Lcd(0x20,16,2);
```

Listing 9-3: *I²C LCD Test*

```
// Listing 9-3
// I2C LCD Test

#include <Wire.h>
#include <LiquidCrystal_I2C.h>

// set the I2C address to 0x20 for PCF8574P
// A0==A1==A2==0 => addr = 0b00100000 = 0x20
// for 16-column, 2-row display
LiquidCrystal_I2C Lcd(0x20, 16, 2);

void setup() {
  Lcd.init(); // initialize display
}
```

```
void loop() {
  Lcd.setCursor(0, 0);
  Lcd.print("Hello - ");
  Lcd.setCursor(0, 1);
  Lcd.print(" I2C ok ");
  while(1) {} // infinite loop
}
```

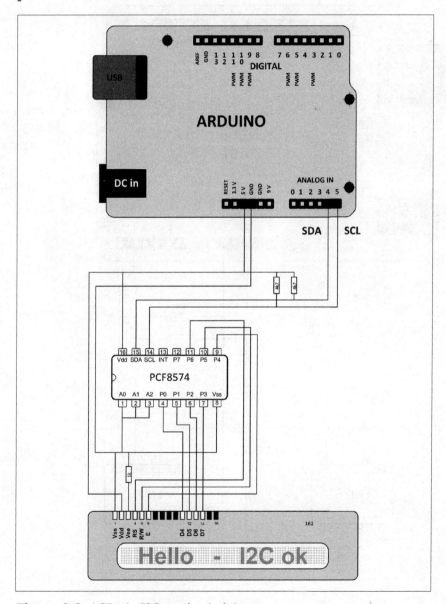

Figure 9.9: *LCD via I²C on the Arduino*

9 INTERFACES

9.6 This Time, Fully Digital: The LM75 Thermometer

The LM75 is another frequently used building block with an I²C interface. It's a complete temperature transducer with an integrated I²C interface. The clear advantage is that the temperature is captured, calculated and digitized by the device and sent to the controller digitally via the I²C.

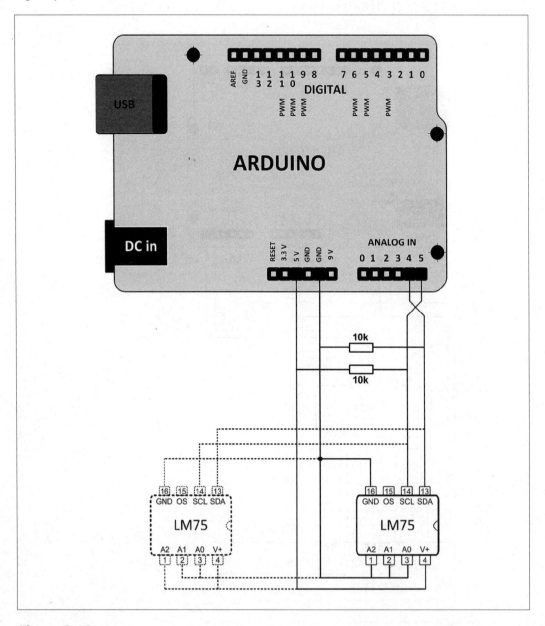

Figure 9.10:
I²C Temperature Sensors on the Arduino. Note the polarity – dimple on the right!

The problems associated with NTCs, such as the nonlinear relationship between temperature and resistance, as well as the requirement of a second resistor to make up a voltage divider, which introduces additional inaccuracy, etc., all fall away.

The LM75 is advantageous even when compared with the analog AD22100. When analog transmission of temperature data takes place over a long cable, errors are easily introduced. Firstly, the intrinsic line impedance may lead to the first errors. Then, electromagnetic radiation, such as from mobile phones and electric motors, may cause unacceptable interference. Lastly, contact stresses at plugs and connections may lead to erroneous readings. For short lines, these influences are usually negligible, but, already for lines in excess of 1 m, problems start occurring in practice. Digital transmission via I²C, on the other hand, is significantly more resilient.

Analog thermometers always require the use of one of the processor's A/D converters and its corresponding pin. For larger applications, this can quickly lead to a pin shortage. Due to the three configurable address pins, up to 8 (2^3) of these sensors may be connected to the I²C bus simultaneously.

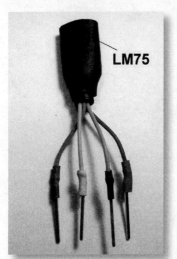

Figure 9.11:
LM75 for Local Measurements

The program starts by including the Wire.h library. Then, for our application, the addresses for two sensors are set. Then, the LCD library is included and the pins defined.

In `setup()`, `Wire` and `Lcd` are initialized. Then follows the printing of a template for the output of temperature values to the LCD. The main loop begins by setting the most significant and least significant bytes for each connected sensor. The two sensors are then read, one after the other. Conversion of the byte values takes place as per the LM75's datasheet. Finally, the temperature reading is output to the LCD.

9 INTERFACES

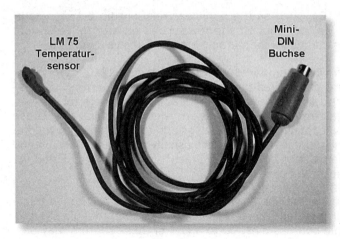

Figure 9.12:
For Use as a Remote Thermometer, a Longer Cable May Be Attached to the LM75.

Listing 9-4: *LM75 LCD Thermometer*

```
// Listing 9-4
// LM75 LCD Thermometer

#include <Wire.h>
#define LM75_1 0b1001000 // LM75 #1 7-bit address
#define LM75_2 0b1001011 // LM75 #2 7-bit address

#include <LiquidCrystal.h>
LiquidCrystal lcd(12, 11, 5, 4, 3, 2); // create LCD object

void setup() {
  Wire.begin();
  lcd.begin(16, 4); // initialize 16x4 LCD
  delay(250);       // for slower modules
  lcd.print("LM75 Thermometer"); delay(1000);
  lcd.setCursor(0, 0); lcd.print("T1 =      C    ");
  lcd.setCursor(0, 1); lcd.print("T2 =      C    ");
}

void loop() {
  byte msb1,lsb1 = 0; byte msb2,lsb2 = 0;
  float degrees1 = 0; float degrees2 = 0;

  // read LM75_1
  Wire.beginTransmission(LM75_1);
  Wire.write(0x00);
  Wire.endTransmission();

  Wire.requestFrom(LM75_1, 2);
  while(2 > Wire.available()) {};

  msb1 = Wire.read();
```

```
    lsb1 = Wire.read();

    if (0x80 > msb1) {
       degrees1 = ((msb1 * 10) + (((lsb1 & 0x80) >> 7) * 5));
    }
    else {
       degrees1 = ((msb1 * 10) + (((lsb1 & 0x80) >> 7) * 5));
       degrees1 = -(2555.0 - degrees1);
    }
    degrees1 /= 10;

    // read LM75_2
    Wire.beginTransmission(LM75_2);
    Wire.write(0x00);
    Wire.endTransmission();

    Wire.requestFrom(LM75_2, 2);

    while (2 > Wire.available()) {};

    msb2 = Wire.read();
    lsb2 = Wire.read();

    if(0x80 > msb2) {
       degrees2 = ((msb2 * 10) + ((( lsb2 & 0x80) >> 7) * 5));
    }
    else {
       degrees2 = ((msb2 * 10) + (((lsb2 & 0x80) >> 7) * 5));
       degrees2 = -(2555.0 - degrees2);
    }
    degrees2 /= 10;

    // write results to LCD
    lcd.setCursor(5, 0); lcd.print(degrees1);
    lcd.setCursor(5, 1); lcd.print(degrees2);
    delay(100);
}
```

9.7 Power-Saving: Real-Time Clock with Date Display

As shown in Chapter 8, a very accurate clock can be made from a microcontroller being clocked by a quartz crystal. In this section, our timekeeping will be further improved. This is achieved by adding a real-time clock (RTC) device, the PCF8583. This contains a high-precision timer that is clocked by a watch crystal. This crystal runs at a relatively low frequency of 32.768 kHz. This number is a power of two ($2^{15} = 32,768$), so, with simple digital integer frequency division, an exact one-second clock signal may be generated. These 32 kHz crystals are very precisely tuned and achieve significantly improved long-term stability compared with the average high-speed microcontroller crystal.

9 INTERFACES

Besides the oscillator, the RTC device contains a complete circuit for calculating and outputting the complete date and time of day. This frees up substantial resources in the main microcontroller, freeing them up for other purposes. The device is connected via the I²C bus, so, again, it is very light on resources.

In addition, the use of a separate RTC device enables the use of a simple backup power supply for the clock. Using a small battery (e.g. 3.6 V, 180 mAh), the device is easily supplied with reserve power in the event of a power failure, allowing the device to continue running for some time. The clock need only be set correctly once, and it will continue to run unfailingly with high precision. A library for running the PCF8583 is available for download at:

github.com/edebill/PCF8583

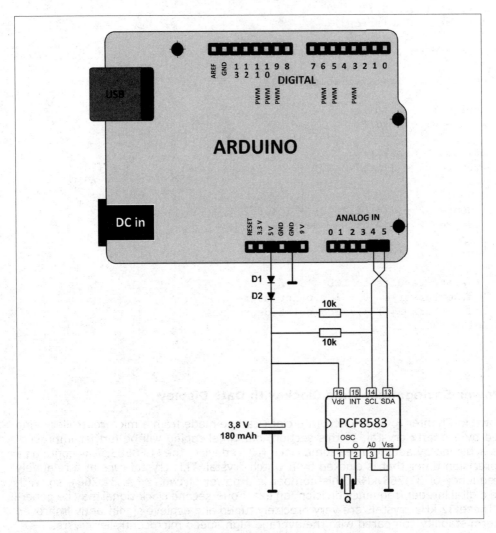

Figure 9.13: *PCF8583 Real-Time Clock Module on Arduino*

POWER-SAVING: REAL-TIME CLOCK WITH DATE DISPLAY 9.7

To set the clock, a string of the form `YYMMddhhmmss;` is sent to the device via the I²C bus. The characters have the expected meanings:

    `YY`: 2-digit year value
    `MM`: 2-digit month value
    `dd`: 2-digit day
    `HH`: 2-digit hour
    `mm`: 2-digit minute
    `ss`: 2-digit second

with a semicolon terminating the string. With

```
131019173000;
```

The time is set to October 19th, 2013, 17:30:00.

Note
    This sketch requires the Arduino-0022 IDE.

In addition to the PCF8583 RTC library, the program also requires that the Wire library be included. The definition of the required variables follows. The I²C address is set to 0, as the single address pin, A0, is tied to ground. The `yearof` variable establishes the millennium. The values `of` and `mu` define two offset constants in the PCF8583 data format.

In `setup()`, in addition to the serial port, an LCD module is initialized. This should be connected in the standard configuration shown in Figure 6.9, in order to display the time data.

The first function block in the main loop performs two tasks. Firstly, it reads the time from the I²C bus. Secondly, it writes the current time and date to the RTC device if this is received via the serial interface.

The rest of the program merely sends the time and date to the serial interface, as well as to the LCD. There are two new features. Firstly, the data is rendered neatly readable by the formatted print function, `sprintf()`. The format `"%02d:%02d:%02d"` serves to display the number sequence in the familiar clock format. This means two decimal digits, with leading zeros. For display on the PC terminal, the data is simply sent in numerical form. The second special feature is that the month value is first converted to an alphabetic representation of the month, so that, for example, instead of 2013-10-19, the display reads 19 October 2013.

Listing 9-5: *Real-Time Clock with LCD*

```
// Listing 9-5
// Real-Time Clock with LCD

// Set time format YYMMddhhmmss;

#include <Wire.h>              // include I2C communication
#include <stdio.h>
```

9 INTERFACES

```
#include <PCF8583.h>                    // include for RTC unit PCF8583
#include <LiquidCrystal.h>
LiquidCrystal Lcd(12, 11, 5, 4, 3, 2); // initialize the LCD interface pins

int correctAddress;
const int yearOf = 2000;
const byte of = 48, mu = 10;
PCF8583 p(0xA0);

void setup(void) {
  Serial.begin(9600);
  Serial.println("Starting...");
  Lcd.begin(16, 4); // 16 columns, 4 rows
  delay(250);       // for slower devices
  Lcd.setCursor(3, 0); Lcd.print("RTC Clock");
}

void loop(void) {
  if (0 < Serial.available()) {
    p.year   = (byte)((Serial.read() - of) * mu + (Serial.read() - of)) + yearOf;
    p.month  = (byte)((Serial.read() - of) * mu + (Serial.read() - of));
    p.day    = (byte)((Serial.read() - of) * mu + (Serial.read() - of));
    p.hour   = (byte)((Serial.read() - of) * mu + (Serial.read() - of));
    p.minute = (byte)((Serial.read() - of) * mu + (Serial.read() - of));
    p.second = (byte)((Serial.read() - of) * mu + (Serial.read() - of));
    if (';' == Serial.read()) {
      Serial.println("Setting date...");
      p.set_time();}
  }

  p.get_time();
  char time[50]; char date[50];
  sprintf(time, "%02d:%02d:%02d", p.hour, p.minute, p.second);
  sprintf(date, "%02d/%02d/%02d", p.year, p.month, p.day);
  Serial.print(date); Serial.print(" - "); Serial.println(time);
  Lcd.setCursor(3, 1); Lcd.print(time);
  Lcd.setCursor(19, 0); Lcd.print(p.day); Lcd.print(". ");
  switch (p.month) {
    case 1 : Lcd.print ("January  "); break;
    case 2 : Lcd.print ("February "); break;
    case 3 : Lcd.print ("March    "); break;
    case 4 : Lcd.print ("April    "); break;
    case 5 : Lcd.print ("May      "); break;
    case 6 : Lcd.print ("June     "); break;
```

```
        case  7 : Lcd.print ("July      "); break;
        case  8 : Lcd.print ("August    "); break;
        case  9 : Lcd.print ("September "); break;
        case 10 : Lcd.print ("October   "); break;
        case 11 : Lcd.print ("November  "); break;
        case 12 : Lcd.print ("December  "); break;
    }

    Lcd.setCursor(21, 1);
    Lcd.print(p.year);
    delay(100);
}
```

9.8 After including the IRremote Wireless, Practical, Quick: The IR Interface

IR (infrared) interfaces are found in many areas of modern consumer electronics. Whether it's a television set, hi-fi center, DVD or Blu-ray player, there is hardly a piece of equipment without an infrared remote control. Even increasingly popular miniature helicopters and tiny model cars are being controlled by infrared remote these days.

With such a widespread use, it makes sense to use infrared remote control to operate custom-made microcontroller-based devices. Virtually every household has at least one IR remote control. There are often so many remote controls in the modern living room that many people are barely able to figure out which one does what. What could be more obvious than using one of these for exciting DIY projects? Plus, the remote control does not need to be modified, so its suitability to its original purpose will not be affected.

Until a few years ago, the Philips RC5 code was widely used in infrared remote controls. This protocol had become a quasi-standard in consumer electronics, but, more recently, the situation has changed: nowadays, almost every manufacture has its own IR code.

In order to use an existing infrared remote for custom applications, it must first be determined which code it uses. For this, the IR library from

arcfn.com/files/IRremote.zip

must be downloaded and installed in the usual manner. Then, by clicking the *Open* icon, under *IRremote*, a test program, *IRrecvDump*, will be available (see Figure 9.14).

Now, all that's left is to connect an IR receiver module to the Arduino. A widely available, often used type is the TWOP1736, while the SFH5110-36 is a good substitute.

These receivers contain not only the receiver diode, but also the entire receiver electronics. These include automatic gain control, a band pass filter and a demodulator. This allows trouble-free operation at decent distances, even in unfavorable light conditions. In addition, the device's apparently black enclosure is actually transparent to infrared light, even though visible light doesn't pass through it. This further increases the receiver module's signal-to-noise ratio.

9 INTERFACES

Figure 9.14:
Test Sketch for IR Remote Controls

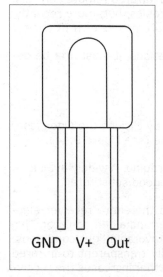

Figure 9.15:
TSOP1736 IR Receiver Pin Assignment

Thanks to the integrated electronics, the TSOP1736's output signal is directly usable by a microcontroller. Connecting the receiver to the Arduino is very easy. Figure 9.16 shows the corresponding schematic.

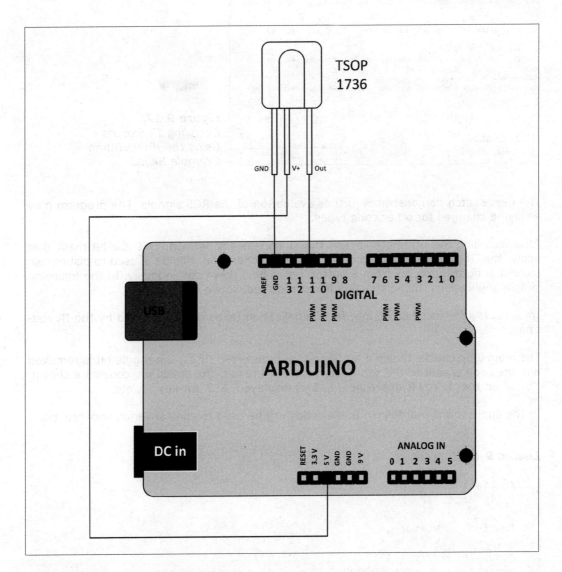

Figure 9.16: TSOP1736 IR Receiver on Arduino

Once the above sketch is uploaded to the Arduino, different IR remotes may be tested. For this, the remote control must be pointed at the TSOP1736. As each button is pressed, its corresponding code will be sent to the serial interface and shown in Serial Monitor. In Figure 9.16, both a Sony and an RC5 remote control are successfully detected.

9 INTERFACES

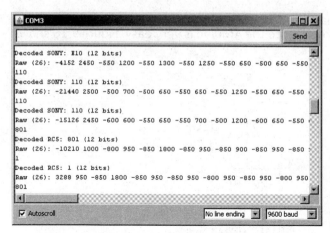

Figure 9.17:
Decoding IR Signals
Using the IRrecvDump
Example Sketch

The next sketch demonstrates further evaluation of the RC5 signals. The program may easily be changed for other code types.

After including the IRremote library, the tB constant is defined. This is a bit mask that allows the 'Toggle' bit to be removed. In normal operation, this bit is used to distinguish individual button presses from a button being held down continuously. In the following simple applications, this distinction is not required, so we mask the bit.

In setup(), the usual serial interface initialization takes place, followed by the IR routines.

The main loop checks to see if an IR code was received. If so, the toggle bit is removed and the code is sent to the serial interface. As a result, the direct key codes are shown, i.e., when the '1' key is depressed, a '1' is displayed, a '2' for key '2', etc.

In the applications that follow, these codes will be used to control various operations.

Listing 9-6: *IR Receiver*

```
// Listing 9-6
// IR Receiver

#include <IRremote.h>

const int irPin = 11;
const long tB = 0b100000000000; // pos of toggle bit
int irCode;

IRrecv IrRecv(irPin);
decode_results results;

void setup() {
  Serial.begin(9600);
```

```
    IrRecv.enableIRIn(); // Start the receiver
}

void loop() {
  if (IrRecv.decode(&results)) {
    irCode = results.value;

    if (tB <= irCode) {
      irCode -= tB; // reduce code
    }

    Serial.println((irCode), DEC);
    IrRecv.resume(); // get next code
  }
}
```

9.9 'Lampino': An IR-Controlled RGB Lamp

Now that RGB LEDs are available at low cost, they are gaining in popularity. One often-encountered application is the so-called mood light, which is a light that automatically cycles through all the colors of the visible spectrum, from red through yellow to green and blue.

In this project, the color of the light won't change automatically, but may be adjusted to suit the user's mood. In this way, green shades may be set for a relaxing ambience, while red is an option for a more romantic time. Blue might indicate a cool, refreshed mood.

The color may be adjusted using a remote control from the comfort of an armchair.

There is one technical clarification. You may wonder why PWM channels 5, 6 and 9 are used, and not one of the other available PWM outputs. The answer is because the IR-remote library is already using Timer 2. The same timer generates the PWM signals for Channels 3 and 11, so these and the IR library may not be used simultaneously.

As shown in Figure 9.18, only three resistors and the RGB LED are required in addition to the IR receiver. The resistor values should be selected so that the individual LEDs are more or less equally bright at maximum voltage. If so, then, when all three PWM channels are set to maximum, a color-neutral white should be the result.

9 INTERFACES

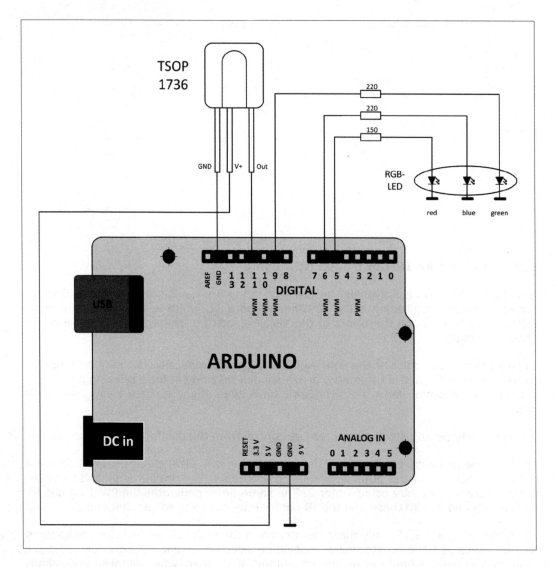

Figure 9.18: *IR-Controlled RGB Lamp*

The accompanying sketch is very simple. Since the program in the previous section already offers us the numerical values, these are simply used to control the individual PWM channels. To do this, the following key assignments are used:

1	2	3	4	5	6
Red +brightness	Green +brightness	Blue +brightness	Red -brightness	Green -brightness	Blue -brightness

Table 15: *Control of the RGB Lamp*

Listing 9-7: *IR-Controlled RGB Lamp*

```
// IR RemCo controlled RGB lamp

#include <IRremote.h>

const int irPin = 11;
const byte ledRed = 9, ledGreen = 5, ledBlue = 6;
const long tB = 0b100000000000; // Toggle Bit position

int irCode;
int pwm1 = 1, pwm2 = 1, pwm3 = 1;
IRrecv IrRecv(irPin);
decode_results results;

void setup() {
  IrRecv.enableIRIn(); // Start the receiver
  pinMode(ledRed, OUTPUT);
  pinMode(ledGreen, OUTPUT);
  pinMode(ledBlue, OUTPUT);
}

void loop() {
  if (IrRecv.decode(&results)) {
    irCode = results.value;
    if (tB <= irCode) {
      irCode -= tB;
    }

    switch (irCode) {
      case 1:
        if (pwm1 < 256) {
          pwm1 *= 2;
        }
        analogWrite(ledRed, pwm1 - 1);
        break;
      case 2:
        if (pwm2 < 256) {
          pwm2 *=2 ;
        }
        analogWrite(ledGreen, pwm2 - 1);
        break;
      case 3:
        if (pwm3 < 256) {
          pwm3 *= 2;
        }
        analogWrite(ledBlue, pwm3 - 1);
        break;
      case 4:
        if (pwm1 > 1) {
          pwm1 /= 2;
```

```
          }
          analogWrite(ledRed, pwm1 - 1);
          break;
        case 5:
          if (pwm2 > 1) {
            pwm2 /= 2;
          }
          analogWrite(ledGreen, pwm2 - 1);
          break;
        case 6:
          if (pwm3 > 1) {
            pwm3 /= 2;
          }
          analogWrite(ledBlue, pwm3 - 1);
          break;
    }
    IrRecv.resume();                              // get next code
  }
}
```

9.10 Timely Luxury: An IR-Controlled Digital Clock

Adding an IR remote control to the digital clock from Section 9.8 gives us a luxury timepiece. Since the operation of IR remotes is now understood, conversion of the clock setting interface from two push-buttons to infrared remote control is quite simple.

With an IR remote control, the clock may be placed in a difficult-to-reach place, such as on top of a high cabinet, and setting the time may be done from just about anywhere in the room. This project also shows that, by combining several techniques, it is possible to create much customized devices that are not easily purchased in a store.

The circuit construction doesn't present any problems. The display unit connection is borrowed from the digital clock in Section 8.2. The only additional requirement is the IR receiver module. If the LED display is placed behind an IR-transparent screen, the IR receiver may also be placed behind the screen, for example right next to the display.

The software can be put together from already-introduced building blocks. Instead of checking for physical key presses, a query is made to check for received IR codes. The software methodology for this is already laid out in the previous sections. The following table shows the RC5 key assignments relevant to the digital clock.

1	2
Hour +	Minute +
4	5
Hour -	Minute -

Table 16: *Key Assignments for the Digital Clock with IR Remote*

TIMELY LUXURY: AN IR-CONTROLLED DIGITAL CLOCK 9.10

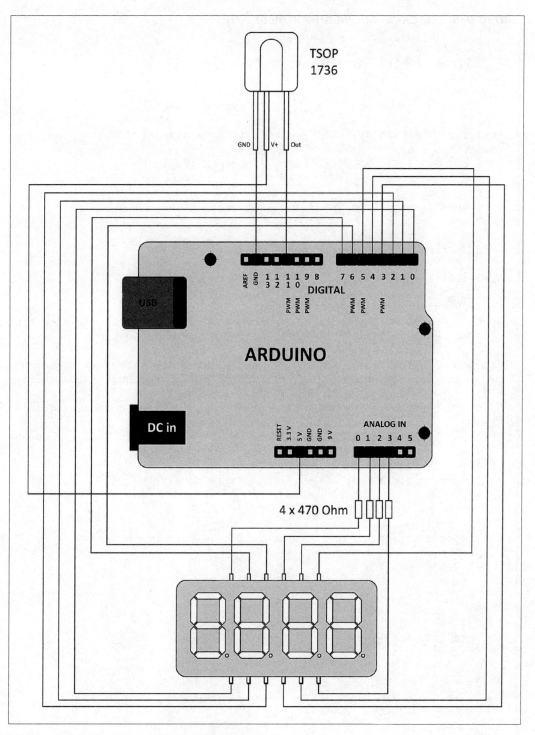

Figure 9.19: IR-Controlled Digital Clock

Listing 9-8: *LED Clock with Infrared Remote*

```
// Listing 9-9
// LED Clock with Infrared Remote

#include <IRremote.h>           // include IR lib
#include <TimerOne.h>           // include timer

#include "LedDisplayV3.h"       // include display driver
const int irPin = 11;
const long tB = 0b100000000000; // pos of toggle bit

int irCode;
int hrs=20, mins=0, secs=0;   // set start time

IRrecv IrRecv(irPin);
decode_results results;

volatile int tc = 0;            // timeCounter

void setup() {
  DDRB = 000000;                // port B (pins digital 8..13) as output
  DDRC = 0b00001111;            // port C 0..3 (pins analog 0..3) as output
  DDRD = 0b11111111;            // port D (pins digital 0..7) as output
  digitalWrite(8, HIGH);        // turn on pullup resistor on B0 (Digital 8)
  digitalWrite(9, HIGH);        // turn on pullup resistor on B1 (Digital 9)
  Timer1.initialize(1000000);   // interrupt every 1000000 us = 1.000000 s
  Timer1.attachInterrupt(update_time);
  IrRecv.enableIRIn();          // start IR receiver
}

void loop() {
  numberOutput (hrs * 100 + mins);
  secs = tc;
  if (60 == secs) {
    tc = 0; secs = 0; mins++;
    if (60 == mins) {
      mins=0; hrs++;
    }
    if (24 == hrs) {
      hrs = 0;
    }
  }
  while (secs == tc) {
    refresh();
  }
```

```
    if (IrRecv.decode(&results)) {
      // set time using IR RemCo
      irCode = results.value;
      if (tB <= irCode) {
        irCode -= tB;
      }
      switch (irCode) {
        case 1:
          if (23 > hrs) {
            hrs++;
          }
          else {
            hrs = 0;
          }
          numberOutput(hrs * 100 + mins);
          break;
        case 4:
          if (0 < hrs) {
            hrs--;
          }
          else {
            hrs = 23;
          }
          numberOutput(hrs * 100 + mins);
          break;
        case 2:
          if (59 > mins) {
            mins++;
          }
          else
          {
            mins = 0;
          }
          numberOutput(hrs * 100 + mins);
          break;
        case 5:
          if (0 < mins) {
            mins--;
          }
          else {
            mins = 59;
          }
          numberOutput(hrs * 100 + mins);
          break;
      }
      IrRecv.resume();                        // get next code
    }
}

void update_time() {
```

        tc++;
    }

9.11 Optimal for Microcontrollers: The PS/2 Interface

PC keyboards and mice with PS/2 interfaces have been almost completely replaced by the USB types in recent years. Most new PCs don't even have PS/2 interfaces, so the PS/2 mouse is hardly used for that purpose, these days. The situation is different for microcontroller applications. Thanks to their simple design and signal structure, PS/2 devices are easily used with microcontrollers.

Another advantage is that devices with PS/2-Interface are available at little to no cost. The fact that they are rarely used in their primary application area anymore means that PS/2 devices are available used from online auction sites at very attractive prices, or for free from friends and acquaintances if they have this type of unwanted hardware lying around.

Aside from ground and the power source, the PS/2 interface requires only two signal wires: Clock and Data. The PS/2 socket pin assignment is shown in Figure 9.20.

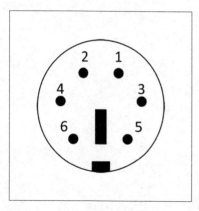

Figure 9.20:
PS/2 Pin Assignment
(Male Connector)

The pins are assigned as follows:
    1) Data
    2) Not connected
    3) Ground
    4) Vcc (+5 V)
    5) Clock
    6) Not connected

Either a mouse or a keyboard may be connected to this socket. Working with the PS/2 signals is again done by exploiting a suitable Arduino library.

9.12 Keyboard and Mouse as Universal Input Devices

In Figure 9.21, the connection of a PS/2 mouse to the Arduino is shown. Again, the connector is shown from the male, or pin side.

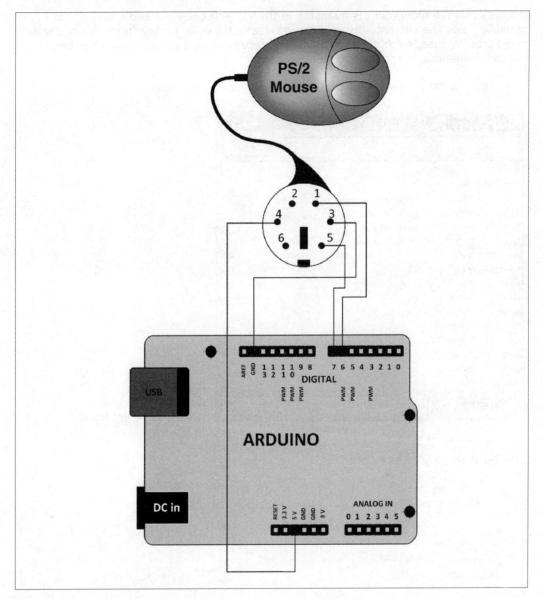

Figure 9.21: *PS/2 Mouse on Arduino*

The PS/2 library may be downloaded from

>   arduino.cc/playground/ComponentLib/Ps2mouse

and installed in the usual way.

In our first application, the functioning of the mouse is tested. After including the library, initialization of the serial interface and mouse in `setup()`, the mouse data is read. The

9 INTERFACES

mouse's *relative* movement is available in the variables `mouseX` and `mouseY`. The `click` variable holds the current status of the mouse buttons, in binary form. For a standard two-button mouse, each of the last two bits represent one of the mouse buttons and is set to '1' whenever a mouse button is pressed.

The captured data is then sent to the PC via the serial interface.

Figure 9.22:
Analysis of mouse signals in Serial Monitor

Listing 9-9: *PS/2 Mouse Control*

```
// Listing 9-9
// PS/2 Mouse Control

#include <ps2.h>

// pin 7: mouse clock; pin 6: mouse data;
PS2 Mouse(7, 6);

char click, mouseX, mouseY;

void setup() {
  Serial.begin(9600);
  mouseInit();
}

// get mouse data and write to PC
void loop() {
  Mouse.write(0xeb);        // get data
```

```
    Mouse.read();            // confirm
    click = Mouse.read();
    mouseX = Mouse.read();
    mouseY = Mouse.read();

    // send data
    Serial.print("X=");
    Serial.print(mouseX, DEC);
    Serial.print("\tY=");
    Serial.print(mouseY, DEC);
    switch (click) {
      case 9:
        Serial.print("\t left click");
        break;
      case 10:
        Serial.print("\t right click");
        break;
      case 11:
        Serial.print("\t both click");
        break;
    }
    Serial.println();
    delay(20);
}

void mouseInit() {
    Mouse.write(0xff);       // reset
    Mouse.read();            // confirm byte 3 times
    Mouse.read();
    Mouse.read();
    Mouse.write(0xf0);       // remote mode
    Mouse.read();            // confirm
    delayMicroseconds(100);
}
```

In our later chapter on physical computing, projects are presented in which various devices are controlled using the mouse.

9.13 A Complete Microcomputer with LCD Monitor and Keyboard

Connect a keyboard and a mouse to the Arduino, and you have yourself a complete microcomputer. Together with an LCD display, it forms a system with performance comparable to the first early microcomputers.

In the following sketch, a simple 'typewriter' is presented as a simple application. For this, the characters received from the keyboard are displayed on the LCD 'monitor'.

Whenever a key is pressed, a keyboard generates a scan code. These scan codes don't correspond to the ASCII table in any simple way. Therefore, it's necessary to fix each

9 INTERFACES

scan code to its corresponding character code in the Arduino sketch, and this task assignment constitutes a major part of the sketch. Aside from this, only the `ps2.h` and `LiquidCrystal.h` libraries are included. After the initialization routines, the main loop simply converts the received scan codes into their ASCII equivalents, where available, and sends them to the connected LCD.

Extensions and exercises

- Add some features to the 'typewriter' sketch to make it a 'word processor', capable of editing, i.e. inserting and deleting.

- Create a code lock that opens only when a certain secret code is entered via the keyboard.

Microcontroller systems with keyboard connections are common in many technical areas. The main applications are of course in the field of machinery and equipment control. With the increasing efficiency of microcontroller systems, it's a small transition to the classic computer system with peripheral devices. Whether a microcontroller or traditional computer is to be used will depend on the application.

Listing 9-10: *PS/2 Keyboard and LCD*

```
// Listing 9-10
// PS/2 Keyboard and LCD

#include <ps2.h>
#include <LiquidCrystal.h>

// Initialize keyboard using Pin 6 = data & pin 7 = clock
PS2 Keyboard(7, 6);

// creat Lcd object, specify pins
LiquidCrystal Lcd(12, 11, 5, 4, 3, 2);

unsigned char code;

void keyboardInit() {
  char ack;
  Keyboard.write(0xff);  // send reset code
  ack = Keyboard.read(); // byte, kbd does self test
  ack = Keyboard.read(); // another ack when self test is done
}

void setup() {
  // Initialize LCD: 16 columns, 4 rows
  Lcd.begin(16, 4);
  // Print a start message to the LCD
  Lcd.print("Starting...");
  delay(1000);
```

```
    Lcd.setCursor(0, 0);
    Lcd.print("              ");
    Lcd.setCursor(0, 0);
    keyboardInit();
}

void loop() {
    // get keycode and send it to the LCD
    // read 3 scan codes per keystroke, discarding the first two
    code = Keyboard.read(); code = Keyboard.read(); code = Keyboard.read();

    // The scan code in each case statement on the left corresponds
    // with the character sent to the Lcd.print() function on the right:
    switch (code) {
        case 0x15: Lcd.print("q"); break;
        case 0x16: Lcd.print("1"); break;
        case 0x1a: Lcd.print("y"); break;
        case 0x1b: Lcd.print("s"); break;
        case 0x1c: Lcd.print("a"); break;
        case 0x1d: Lcd.print("w"); break;
        case 0x1e: Lcd.print("2"); break;
        case 0x21: Lcd.print("c"); break;
        case 0x22: Lcd.print("x"); break;
        case 0x23: Lcd.print("d"); break;
        case 0x24: Lcd.print("e"); break;
        case 0x25: Lcd.print("4"); break;
        case 0x26: Lcd.print("3"); break;
        case 0x29: Lcd.print(" "); break;
        case 0x2a: Lcd.print("v"); break;
        case 0x2b: Lcd.print("f"); break;
        case 0x2c: Lcd.print("t"); break;
        case 0x2d: Lcd.print("r"); break;
        case 0x2e: Lcd.print("5"); break;
        case 0x31: Lcd.print("n"); break;
        case 0x32: Lcd.print("b"); break;
        case 0x33: Lcd.print("h"); break;
        case 0x34: Lcd.print("g"); break;
        case 0x35: Lcd.print("y"); break;
        case 0x36: Lcd.print("6"); break;
        case 0x39: Lcd.print(","); break;
        case 0x3a: Lcd.print("m"); break;
        case 0x3b: Lcd.print("j"); break;
        case 0x3c: Lcd.print("u"); break;
        case 0x3d: Lcd.print("7"); break;
        case 0x3e: Lcd.print("8"); break;
        case 0x41: Lcd.print(","); break;
        case 0x42: Lcd.print("k"); break;
        case 0x43: Lcd.print("i"); break;
        case 0x44: Lcd.print("o"); break;
        case 0x45: Lcd.print("0"); break;
```

```
            case 0x46: Lcd.print("9"); break;
            case 0x49: Lcd.print("."); break;
            case 0x4a: Lcd.print("-"); break;
            case 0x4b: Lcd.print("1"); break;
            case 0x4c: Lcd.print(" "); break;
            case 0x4d: Lcd.print("p"); break;
            case 0x4e: Lcd.print("+"); break;
            case 0x61: Lcd.print("<"); break;
            case 0x69: Lcd.print("1"); break;
            case 0x6b: Lcd.print("4"); break;
            case 0x6c: Lcd.print("7"); break;
            case 0x70: Lcd.print("0"); break;
            case 0x71: Lcd.print(","); break;
            case 0x72: Lcd.print("2"); break;
            case 0x73: Lcd.print("5"); break;
            case 0x74: Lcd.print("6"); break;
            case 0x75: Lcd.print("8"); break;
            case 0x79: Lcd.print("+"); break;
            case 0x7a: Lcd.print("3"); break;
            case 0x7b: Lcd.print("-"); break;
            case 0x7c: Lcd.print("*"); break;
            case 0x7d: Lcd.print("9"); break;
        }
    }
```

10
Sounds and Synthesizer

In addition to optical output via LEDs, LCDs or 7-segment displays, microcontrollers are well suited to producing acoustic signals. The best known examples of these are short tones used as warning signals. Simple acoustic sounds are very easy for a microcontroller to produce, as the frequency of human hearing ranges from 16 Hz to around 16 kHz, and these frequencies are no challenge for a micro.

In addition, complex sound effects that create striking impressions are possible. The spectrum ranges from simple electronic organ sounds to sophisticated synthesizers. In principle, even electronic speech synthesis is possible with the Arduino. While the results thereof may not be completely satisfactory, it can be assumed that a lot of progress will be made in this area in the future.

10.1 Simple Tones

The easiest method of creating a tone is by using the `tone()` function, which creates basic square waves that are well suited to the beeps and short melodies familiar from microwave ovens, simple electronic alarms, tumble dryers, etc.

To create these sounds, a piezo transducer is more than adequate. However, an amplifier such as the one from Section 10.1 may be required for higher volumes.

The following program illustrates the generation of a simple interval alarm sound.

Listing 10-1: *Simple Alarm Sound*

```
// Listing 10-1
// Simple Alarm Sound
const int speaker = 9;

void setup() {
  pinMode(speaker, OUTPUT);
}

void loop() {
  tone(speaker, 550, 450);
  delay(1000);
}
```

10 SOUNDS AND SYNTHESIZER

The corresponding construction is shown in the following figure.

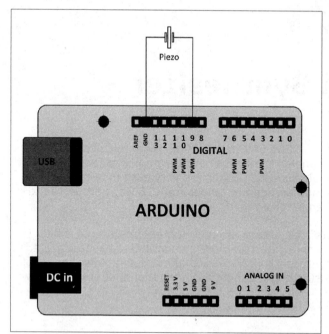

Figure 10.1:
Simple Alarm Tone Generator

More complex sounds may be readily generated using the `tone()` function, as the next example sketch shows.

Listing 10-2: *Red Alert*

```
// Listing 10-2
// Red Alert

const int speaker = 9; // select speaker pin

void setup() {
  pinMode(speaker, OUTPUT);
}

void loop() {
  for (int i = 200; i < 500; i += 10) {
    tone(speaker, i, 50);
    delay(20);
  }
  delay(1000);
}
```

Similar tones are often used in science fiction. You may even be familiar with the sound produced here.

Extensions and exercises

- With the Alarm Sound and Red Alert sketches as a starting point, try to recreate sounds from the world of sci-fi as faithfully as possible.

10.2 Transducers and Amplifiers

The simplest way to get sound out of a microcontroller is to connect a piezoelectric transducer to a processor pin, but piezos aren't very loud, and their sound quality leaves much to be desired. Piezos are therefore generally only used for creating simple beeps.

Much better audio quality is achieved using a dynamic loudspeaker. The problem is that one of these has an impedance of just a few ohms, so it can't be connected directly to an Arduino, as it would cause an overload. Rather, an amplifier is needed for this purpose. A superbly suitable device for this purpose, the LM386, is used in our applications. To create a complete amplifier, only a few additional components are required. The corresponding circuit diagram is shown in Figure 10.2.

In addition to the actual amplifier, an audio filter is created. The 2.2 µF capacitor takes care of the DC decoupling of the amplifier, and also has a certain high-pass characteristic, while the 1 µF capacitor contributes some low-pass filtering. This is particularly desirable when audio signals are generated using simple pulse width modulation. The high PWM harmonics will be well suppressed, while the desired audio frequencies in the lower kHz range are passed on to the amplifier.

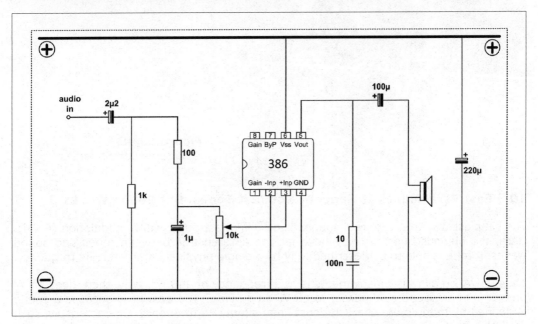

Figure 10.2: Circuit Diagram for an LM386-Based Audio Amplifier

10 SOUNDS AND SYNTHESIZER

This amplifier will fit comfortably on a small breadboard. Figure 10.3 shows a corresponding construction example. If the amplifier is to be deployed more permanently, stripboard may be used.

Figure 10.3:
Construction Example for the Audio Amplifier

Figure 10.4:
A Suitable DIY Speaker Enclosure

10.3 Fast PWM Makes It Happen: Not Just Tones, But Sound Waves

The simplest way to create real sounds is by using fast pulse-width modulation (PWM). Here, the obvious question may be what the difference is between tones and sound waves. A tone is able to be described fully by a single physical parameter: its frequency.

Thus, an 'A' on all music instruments is at a frequency of 440 Hz. Why, then, does an 'A' on a trombone sound so different to an 'A' on a piano? The answer is because a sound is defined by many more characteristics than simply its frequency. Characteristics such as timbre, defined by the mix of harmonic overtones, as well as the temporal variation in amplitude, or so-called envelope, are significant.

The `tone()` function allows no influence on these variables. It creates a simple, fixed-fre-

FAST PWM MAKES IT HAPPEN: NOT JUST TONES, BUT SOUND WAVES 10.3

quency square wave for a defined period of time. Because square waves have many harmonics, they sound harsh and technical. For smoother sounds, these harmonics must be suppressed. For a soft and harmonious sound, pure sine waves are the order of the day. These cannot be produced by simply turning a port pin on and off. One option is to use pulse-width modulation. With this method, a square wave is produced with a frequency way above the human hearing range, e.g. 62.5 kHz. The frequency remains constant, but the on:off timing ratio is varied. With the appropriate analog filtering, an almost continuously varying signal may be produced. This is a quasi-analog output that may, in principle, be capable of approximating any wave form.

The following program generates a roughly sinusoidal waveform.

Listing 10-3: *Fast PWM Sine Wave*

```
// Listing 10-3
// Fast PWM Sine Wave

const int audioPin = 9; // audio output pin

const byte value[] = {
  128, 131, 134, 137, 141, 144, 147, 150, 153, 156, 159, 162, 165,
168, 171, 174,
  177, 180, 183, 186, 189, 191, 194, 197, 199, 202, 205, 207, 209,
212, 214, 217,
  219, 221, 223, 225, 227, 229, 231, 233, 235, 236, 238, 240, 241,
243, 244, 245,
  246, 247, 248, 249, 250, 251, 252, 253, 253, 254, 254, 255, 255,
255, 255, 255,
  255, 255, 255, 255, 255, 254, 254, 254, 253, 253, 252, 251, 250,
249, 248, 247,
  246, 245, 243, 242, 240, 239, 237, 236, 234, 232, 230, 228, 226,
224, 222, 220,
  218, 215, 213, 211, 208, 206, 203, 201, 198, 195, 193, 190, 187,
184, 181, 179,
  176, 173, 170, 167, 164, 161, 158, 155, 152, 148, 145, 142, 139,
136, 133, 130,
  126, 123, 120, 117, 114, 111, 108, 104, 101, 98, 95, 92, 89, 86,
83, 80,
  77, 75, 72, 69, 66, 63, 61, 58, 55, 53, 50, 48, 45, 43, 41, 38,
  36, 34, 32, 30, 28, 26, 24, 22, 20, 19, 17, 16, 14, 13, 11, 10,
  9, 8, 7, 6, 5, 4, 3, 3, 2, 2, 1, 1, 0, 0, 0, 0,
  0, 0, 0, 1, 1, 1, 2, 2, 3, 4, 4, 5, 6, 7, 8, 10,
  11, 12, 13, 15, 16, 18, 20, 21, 23, 25, 27, 29, 31, 33, 35, 37,
  39, 42, 44, 47, 49, 51, 54, 57, 59, 62, 65, 67, 70, 73, 76, 79,
  82, 85, 88, 91, 94, 97, 101, 103, 106, 109, 112, 115, 119, 122, 125
};

void setup() {
  pinMode(audioPin, OUTPUT);
  TCCR1A = 0b10000001;
```

10 SOUNDS AND SYNTHESIZER

```
    TCCR1B = 0b00001001;
}

void loop() {
  for (unsigned int j = 0; j < 256; j++) {
    analogOut(value[j]);
    delayMicroseconds(10);
  }
}

void analogOut(byte val) {
  OCR1A = (val);
}
```

For the sake of processing speed, we used the technique of pre-calculating the sine values and storing them in the `value[]` array. In the main loop, the individual amplitude values are then simply output using the fast PWM function.

The two lines

```
TCCR1A = 0b10000001;
TCCR1B = 0b00001001;
```

set the PWM frequency to 62.5 kHz. These instructions place the required values directly in the processor's Timer and Counter Control Registers (TCCR). A detailed description of this technique is beyond the scope of this book, but for further details, consult the chapter on Timers, Counters and PWM in the ATmega328P datasheet.

Here again, we see the advantages of using the Arduino programming environment. Since the compiler is already using the AVR-GCC compiler in the background, it's easy to include some advanced code. Had we used a special programming language just for beginners, we'd have come up against some difficult limitations.

Extensions and exercises

- Alternately load the two sketches, *Fast PWM Sine Wave* and *Alarm*, on the Arduino, and compare their sounds.

The following sketch allows a direct auditory comparison between different waveforms. For this, a square wave is first generated, followed by a sawtooth and then a sine wave. Finally, sine waves of differing amplitudes are generated. This demonstrates control of not only frequency and harmonics, but amplitude as well.

FAST PWM MAKES IT HAPPEN: NOT JUST TONES, BUT SOUND WAVES 10.3

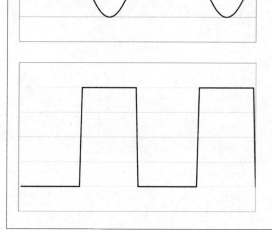
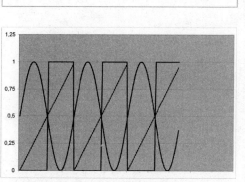
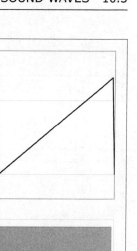

Figure 10.5: Sine, Sawtooth and Square Waves of the Same Frequency

Listing 10-4: Fast PWM Waveforms

```
// Listing 10-4
// Fast PWM Waveforms

const int audioPin = 9;                      // audio output for fast PWM

const byte value[] = {
  128, 131, 134, 137, 141, 144, 147, 150, 153, 156, 159, 162, 165,
168, 171, 174,
  177, 180, 183, 186, 189, 191, 194, 197, 199, 202, 205, 207, 209,
212, 214, 217,
  219, 221, 223, 225, 227, 229, 231, 233, 235, 236, 238, 240, 241,
243, 244, 245,
  246, 247, 248, 249, 250, 251, 252, 253, 253, 254, 254, 255, 255,
255, 255, 255,
  255, 255, 255, 255, 255, 254, 254, 254, 253, 253, 252, 251, 250,
249, 248, 247,
  246, 245, 243, 242, 240, 239, 237, 236, 234, 232, 230, 228, 226,
224, 222, 220,
  218, 215, 213, 211, 208, 206, 203, 201, 198, 195, 193, 190, 187,
184, 181, 179,
  176, 173, 170, 167, 164, 161, 158, 155, 152, 148, 145, 142, 139,
```

```
    136, 133, 130,
      126, 123, 120, 117, 114, 111, 108, 104, 101, 98, 95, 92, 89, 86,
    83, 80,
      77, 75, 72, 69, 66, 63, 61, 58, 55, 53, 50, 48, 45, 43, 41, 38,
      36, 34, 32, 30, 28, 26, 24, 22, 20, 19, 17, 16, 14, 13, 11, 10,
      9, 8, 7, 6, 5, 4, 3, 3, 2, 2, 1, 1, 0, 0, 0, 0,
      0, 0, 0, 1, 1, 1, 2, 2, 3, 4, 4, 5, 6, 7, 8, 10,
      11, 12, 13, 15, 16, 18, 20, 21, 23, 25, 27, 29, 31, 33, 35, 37,
      39, 42, 44, 47, 49, 51, 54, 57, 59, 62, 65, 67, 70, 73, 76, 79,
      82, 85, 88, 91, 94, 97, 101, 103, 106, 109, 112, 115, 119, 122, 125
};

void setup() {
  pinMode(audioPin, OUTPUT);
  TCCR1A = 0b10000001;
  TCCR1B = 0b00001001;
}

void loop() {
  // square
  for (unsigned int d = 0; d < 3; d++) {
    for (unsigned int j = 0; j < 100; j++) {
      analogOut(0);
      delayMicroseconds(3000);
      analogOut(255);
      delayMicroseconds(3000);
    }
  }
  delay(1000);

  // sawtooth
  for (unsigned int d = 0; d < 500; d++) {
    for (unsigned int j = 0; j < 255; j++) {
      analogOut(j);
      delayMicroseconds(10);
    }
  }
  delay(1000);

  // sine
  for (unsigned int d = 0; d < 1000; d++) {
    for (unsigned int j = 0; j < 256; j++) {
      analogOut(value[j]);
      delayMicroseconds(10);
    }
  }
  delay(1000);

  // volume variation
  for (int ampl = 1; ampl < 10; ampl++) {
    for (unsigned int d = 0; d < 100; d++) {
```

FAST PWM MAKES IT HAPPEN: NOT JUST TONES, BUT SOUND WAVES   10.3

```
      for (unsigned int j = 0; j < 256; j++) {
        analogOut(value[j]/10*ampl);
        delayMicroseconds(10);
      }
    }
  }

  for (int ampl = 10; ampl >= 0 ; ampl--) {
    for (unsigned int d = 0; d < 100; d++) {
      for (unsigned int j = 0; j < 256; j++) {
        analogOut(value[j] / 10 * ampl);
        delayMicroseconds(10);
      }
    }
  }
  delay(1000);
}

void analogOut(byte val) {
  OCR1A = (val);
}
```

In the next sketch, bell-like sounds are produced. This impression is created by the slow decay of the sine wave's amplitude. This is achieved using a nested loop. While the inner loop creates the fundamental frequency, the outer loop drops the amplitude slowly to zero.

You may be familiar with sounds like this one. They are often heard in cars, airplanes or public transport to draw attention to particular events. Typical examples are the warning signal when a seat belt is not engaged, or if the outdoor temperature is below 3 °C and one should be wary of ice danger.

Listing 10-5: *Fast PWM Bell Sound*

```
// Listing 10-5
// Fast PWM Bell sound

const int audioPin = 9;

byte value[] = {
  128, 131, 134, 137, 141, 144, 147, 150, 153, 156, 159, 162, 165,
168, 171, 174,
  177, 180, 183, 186, 189, 191, 194, 197, 199, 202, 205, 207, 209,
212, 214, 217,
  219, 221, 223, 225, 227, 229, 231, 233, 235, 236, 238, 240, 241,
243, 244, 245,
  246, 247, 248, 249, 250, 251, 252, 253, 253, 254, 254, 255, 255,
255, 255, 255,
  255, 255, 255, 255, 255, 254, 254, 254, 253, 253, 252, 251, 250,
```

```
    249, 248, 247,
      246, 245, 243, 242, 240, 239, 237, 236, 234, 232, 230, 228, 226,
    224, 222, 220,
      218, 215, 213, 211, 208, 206, 203, 201, 198, 195, 193, 190, 187,
    184, 181, 179,
      176, 173, 170, 167, 164, 161, 158, 155, 152, 148, 145, 142, 139,
    136, 133, 130,
      126, 123, 120, 117, 114, 111, 108, 104, 101, 98, 95, 92, 89, 86,
    83, 80,
      77, 75, 72, 69, 66, 63, 61, 58, 55, 53, 50, 48, 45, 43, 41, 38,
      36, 34, 32, 30, 28, 26, 24, 22, 20, 19, 17, 16, 14, 13, 11, 10,
      9, 8, 7, 6, 5, 4, 3, 3, 2, 2, 1, 1, 0, 0, 0, 0,
      0, 0, 0, 1, 1, 1, 2, 2, 3, 4, 4, 5, 6, 7, 8, 10,
      11, 12, 13, 15, 16, 18, 20, 21, 23, 25, 27, 29, 31, 33, 35, 37,
      39, 42, 44, 47, 49, 51, 54, 57, 59, 62, 65, 67, 70, 73, 76, 79,
      82, 85, 88, 91, 94, 97, 101, 103, 106, 109, 112, 115, 119, 122, 125
};

void setup() {
  pinMode(audioPin, OUTPUT);
  TCCR1A = 0b10000001;
  TCCR1B = 0b00001001;
}

void loop() {
  for (int ampl = 10; ampl >= 0 ; ampl--) {
    for (unsigned int d = 0; d < 40; d++) {
      for (unsigned int j = 0; j < 256; j++) {
        analogOut(value[j] / 10 * ampl);
        delayMicroseconds(5);
      }
    }
  }
  delay(100);

  for (int ampl = 10; ampl >= 0 ; ampl--) {
    for (unsigned int d = 0; d < 40; d++) {
      for (unsigned int j = 0; j < 256; j++) {
        analogOut(value[j]/10*ampl);
        delayMicroseconds(5);
      }
    }
  }
  delay(100);

  for (int ampl = 10; ampl >= 0 ; ampl--) {
    for (unsigned int d = 0; d < 20; d++) {
      for (unsigned int j = 0; j < 256; j++) {
        analogOut(value[j] /10 * ampl);
        delayMicroseconds(10);
      }
```

```
      }
    }
    delay(300);
}

void analogOut(byte val) {
  OCR1A = (val);
}
```

Extensions and exercises

- In real percussion instruments, such as bells, triangles, drums and cymbals, the sound doesn't decay linearly, but exponentially. Can you imitate these sounds convincingly?

10.4 Theremin: The Contactless Musical Instrument

Now that our Arduino is creating different tones and sounds, a complete musical instrument will be created. To get by with as few components as possible and simultaneously take advantage of the Arduino's capabilities, it will be a very special instrument. The first example of such a device was developed and presented by Leon Theremin in 1919. It was, of course, completely analog. Oscillators producing the sounds were capacitively detuned using antennas. On the Arduino, capacitive sensors will likewise be used. For this, the CapSense library must be installed. Refer to the details thereon in Section 3.4.

A capacitive sensor consists of a 10 MΩ resistor and a spiral antenna. The spiral antenna is easily made from a length of insulated wire. 10 MΩ resistors are available at most electronics suppliers. If they prove difficult to find, two 4.7 MΩ resistors in series will do the trick (see also Section 7.5).

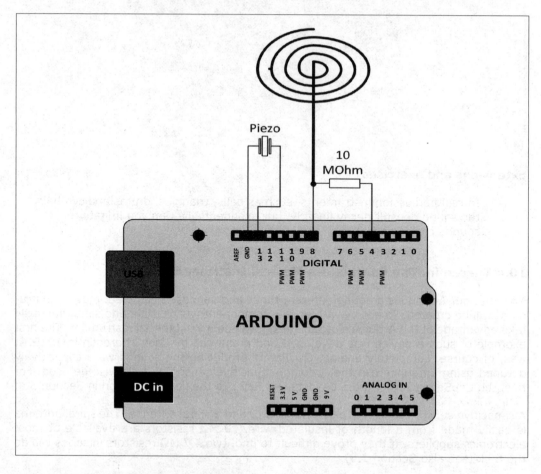

Figure 10.6: Theremin Circuit Diagram

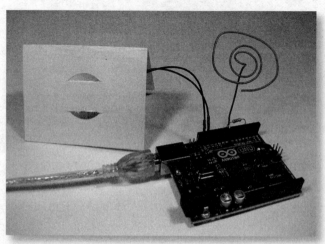

Figure 10.7: Theremin Construction

Listing 10-6: Theremin

```
// Listing 10-6
// Theremin

#include <CapSense.h>

CapSense Cs = CapSense(4,8); // 10 megaohm is between Pins 4 & 8
// Pin 8 is sensor pin
const byte speaker = 13;                // audio output channel
const int len = 300, del = 20;          // tone length and delay

// tone frequencies
const int A = 440, H = 494;
const int C = 523, d = 587, e = 659, f = 698;
const int g = 784, a = 880, h = 988, c = 1047;
const byte threshold = 30; // minimum required sensor output

void setup() {
  pinMode(13, OUTPUT);
  Serial.begin(9600);
}

void loop() {
  long v = Cs.capSense(30);
  Serial.println(v); // send sensor value to PC
  if (threshold < v) {
    digitalWrite(13, HIGH);
    if (v < 40) { tone(speaker, A, 500); }
    if ((v >= 40)  & (v < 50))  { tone(speaker, H, len); }
    if ((v >= 50)  & (v < 60))  { tone(speaker, c, len); }
    if ((v >= 60)  & (v < 70))  { tone(speaker, d, len); }
    if ((v >= 70)  & (v < 80))  { tone(speaker, e, len); }
    if ((v >= 80)  & (v < 90))  { tone(speaker, f, len); }
    if ((v >= 90)  & (v < 100)) { tone(speaker, g, len); }
    if ((v >= 100) & (v < 110)) { tone(speaker, a, len); }
    if ((v >= 120) & (v < 130)) { tone(speaker, h, len); }
    if  (v >= 130)              { tone(speaker, c, 500); }
  }
  else digitalWrite(13, LOW);
  delay(del);
}
```

10.5 Audio Processing

In this chapter, we will see that digital signal processing is not the exclusive domain of digital signal processors (DSPs). The Arduino has sufficient processing power to process audio signals in real time. In this section, the examples begin with tones and sounds, and progress to audio input into the Arduino. In this way, familiar effects such as phaser, reverb, flange, and ring modulation are all possible. Direct generation can produce a versatile range of sounds, from simple sine waves to bell sounds to a xylophone simulation.

In order to feed audio signals into the Arduino, signal matching is required. Audio signals from familiar sources, such as line outputs, headphone jacks or preamps are bipolar alternating voltages. The Arduino's analog inputs only handle voltages of between 0 and +5 V, however, so, it's necessary to add a 2.5 V offset voltage to the signal. Two resistors provide for this (see Figure 10.8). The waveform may then be coupled via a capacitor.

On the output side, the amplifier from Section 10.1 is used. Alternatively, simple active speakers may be used.

To generate test signals, software called Audacity may be used. It is available for free download from

audacity.sourceforge.net/

With this application, audio signals may be generated and processed. All audio signals fed to the Arduino should go through a 3 kHz low-pass filter to avoid aliasing. This low-pass function can be created directly in Audacity. Aliasing effects occur when the ADC sampling frequency is not at least twice as high as the maximum frequency in the supplied signal. The requirement that the sampling rate of the converter be at least twice as high as the signal bandwidth to be processed is also known as the Nyquist rule. If this criterion is not met, higher frequency signal components are 'mirrored' into the used frequency range, creating undesirable effects.

Signal level adjustment is best done using the PC or other playback device's volume control. The volume usually has to be quite high, because the Arduino's ADC requires ±2.5 V in order to be fully driven. For audio signals, this is a fairly high level. If microphones or other sources with relatively low output levels are connected, a preamplifier is advisable.

10.5.1 VCO: A Tunable Sine Wave Source

This voltage-controlled oscillator application impressively demonstrates that the Arduino is capable of producing not only digital signals, but high quality waveforms. The output signal is a continuous sine wave that may be adjusted within a wide frequency range.

The signal frequency is set by the input voltage at A0. For test purposes, a 10 kΩ potentiometer may be used. The result is a very useful test sine generator that may find application in many areas of the audio field. The tuning range is from about 30 Hz to over 5 kHz.

The sketch contains two functions. In the first, `void calcSine()`, calculation of a sine table takes place. This time, the values are not externally pre-calculated, but calculated by the microcontroller itself once and then saved. The second function is called every 16 ms by an automatic interrupt occurring at a frequency of 62.5 Hz. At every fourth call of this interrupt function, the value at analog input A0 is captured and used in the main loop to set the frequency of the sine wave.

In `setup()`, we ensure that Timer 2 is set to Fast PWM mode, and that an interrupt is triggered at the above-mentioned time interval. For this, an individual ATmega328 register is accessed directly. Again, a detailed explanation of these registers is again beyond this book's scope. Consult the Atmel datasheet for interest. At this point, another advantage of the Arduino programming environment becomes clear: it's possible to include standard C code in a sketch. This makes the transition from the Arduino language to professional programming seamless.

The VCO (voltage-controlled oscillator) is definitely one of the most important elements in electronic synthesizers. It is used to create fundamental musical tones, which are turned into familiar electronically synthesized sounds by using filters, mixers and envelopes.

Listing 10-7: *Voltage-Controlled Oscillator*

```
// Listing 10-7
// Voltage-Controlled Oscillator

#define sbi(PORT, bit) (PORT |= 1 << bit)
#define cbi(PORT, bit) (PORT &= ~(1 << bit))

const int pwmOutPin = 11;     // PWM output
const int testPin = 3;
const float pi = 3.141592;

int indexCnt, interCnt; // index, interrupt counters
byte sine[512];              // array for sine values

// interrupt control variables
volatile boolean adcSample;
volatile byte bufferAdc0;

void setup() {
  pinMode(pwmOutPin, OUTPUT);  // sets the PWM pin as output
  pinMode(testPin, OUTPUT);    // sets the test pin as output
  calcSine();                  // load memory with sine table

  // set adc prescaler to 64, 8-Bit ADC and VCC Reference
  // select analog channel 0
  ADCSRA |= 0b00000011;
  ADMUX |= 0b01100000;

  // set Timer2 to fast PWM mode
  TCCR2A |= 0b10000011;

  // Timer2 no prescaler
  TCCR2B |= 0b00000001;
  cbi(TCCR2B, CS21);
  cbi(TCCR2B, CS22);

  // disable Timer0, enable Timer2 interrupt
  cbi(TIMSK0,TOIE0);
```

```
    sbi(TIMSK2,TOIE2);
}

void loop() {
  while (!adcSample) {} // wait for new ADC samples
  digitalWrite(testPin, HIGH);
  adcSample = false;
  OCR2A = sine[indexCnt];   // send current value to PWM
  indexCnt++;
  indexCnt += bufferAdc0;
  indexCnt &= 511;          // constraint counter index to between 0 and 511
  digitalWrite(testPin, LOW);
}

//calculate sine array with 512 values between 0 and 2 * pi
void calcSine() {
  for (int n = 0; n <= 511; n++) {
    sine[n] = 127 * sin(pi * n / 255) + 128;
  }
}

// get potentiometer value
ISR(TIMER2_OVF_vect) {
  interCnt++; // reduce sapmling rate 62.5 kHz through 15.625 kHz
  if (4 == interCnt) {
    bufferAdc0 = ADCH; // get 8-bit ADC value
    adcSample = true;
    sbi(ADCSRA, ADSC); // start next conversion
    interCnt = 0;
  }
}
```

10.5.2 Digital Signal Processing

Aside from generating high-quality analog signals, the Arduino may also be used as a signal processor. In this section, a reverb generator is presented as an example.

In analog technology, the production of artificial reverberation has always presented a particular challenge. While high-, low- and band-pass filters were relatively easy to create, generating reverb was difficult. Mechanical devices such as reverberation plates or springs were tried, and, later, bucket brigade storage was employed. However, these all made for some complicated circuitry.

In a microcontroller, the reverb effect may be realized using a circular memory buffer. For this, the analog signal is first digitized, as usual, via the ADC. Again, the Arduino's ADC operates only on input voltages between 0 and +5 V, so a 2.5 V offset voltage must be added to the input signal. A coupling capacitor sends the adjusted signal in the range of between 0 and 5 V to the ADC input. Figure 10.8 shows a simple circuit for this purpose.

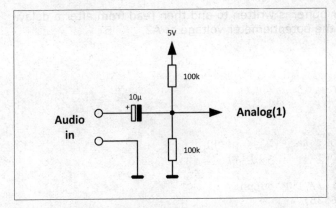

Figure 10.8:
Input Circuit for Reverb Generator

The digitized values are then stored in an array. To produce the delay required by the reverb effect, a circular buffer is read repeatedly over time. The delay is adjusted using a separate potentiometer.

Due to the ATmega328's limited resources, we've limited the array to 512 bytes, but this is sufficient to provide a suitable demonstration.

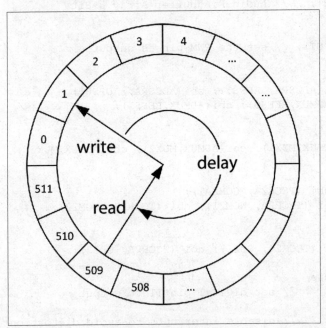

Figure 10.9:
Circular Buffer

Just as in the sine wave generator program, the Fast PWM mode is used. The related settings are again found in `setup()`.

This time, both the potentiometer value and the analog input signal are sampled by the interrupt routine. For this, A1 (analog input) and A2 (potentiometer) are read out alternately.

10 SOUNDS AND SYNTHESIZER

In the main program, the circular buffer is written to and then read from after a delay. The time delay is proportional to the potentiometer voltage at A2.

Listing 10-8: *Reverb Generator*

```
// Listing 10-8
// Reverb Generator

#define cbi(PORT, bit) (PORT &= ~(1 << bit))
#define sbi(PORT, bit) (PORT |= 1 << bit)

const int pwmOutPin = 11; // PWM output
const float pi = 3.1415926535;
const byte halfMax = 127;
boolean div2, selChan;  // control variables for sampling

// interrupt variables accessed globally
volatile boolean adcSample;
volatile byte potSample, analogSample;
int indexCounter, valAnalog, valPot;
byte pwmBuf, ringbuff[512];   // Audio Ringbuffer Array 8-Bit

void setup() {
  pinMode(pwmOutPin, OUTPUT); // set the PWM pin as output

  // adjust adc prescaler
  cbi(ADCSRA, ADPS2); sbi(ADCSRA, ADPS1); sbi(ADCSRA, ADPS0);
  sbi(ADMUX,ADLAR); sbi(ADMUX,REFS0); cbi(ADMUX,REFS1);

  // Set muxer ADC0
  cbi(ADMUX,MUX0); cbi(ADMUX,MUX1); cbi(ADMUX,MUX2); cbi(ADMUX,MUX3);

  // fast PWM for Timer2
  cbi (TCCR2A, COM2A0); sbi (TCCR2A, COM2A1);
  sbi (TCCR2A, WGM20); sbi (TCCR2A, WGM21); cbi (TCCR2B, WGM22);

  // set Timer2 Prescaler
  sbi (TCCR2B, CS20); cbi (TCCR2B, CS21); cbi (TCCR2B, CS22);

  // Timer2 PWM Port Enable
  sbi(DDRB,3);                // set digital pin 11 to output

  //cli();                    // disable interrupts to avoid distortion
  cbi (TIMSK0,TOIE0);         // disable Timer0!!! delay is off now
  sbi (TIMSK2,TOIE2);         // enable Timer2 Interrupt

  valAnalog = analogSample;
}

void loop() {
```

```
    while (!adcSample) {}
    adcSample = false;

    // scale delayed sample with pot value
    valPot = (halfMax - ringbuff[indexCounter]) * potSample / 255;
    valAnalog = halfMax - analogSample;
    valAnalog = valAnalog + valPot;

    // limit analov value (min/max)
    if (valAnalog < -halfMax) valAnalog = -halfMax;
    if (valAnalog > halfMax) valAnalog = halfMax;

    // add offset to PWM buffer
    pwmBuf = halfMax + valAnalog;
    ringbuff[indexCounter] = pwmBuf;
    indexCounter++;

    // reduce bufferindex to 0..511 intervall
    indexCounter = indexCounter & 511;

    OCR2A = pwmBuf; // write current sample to PWM channel
}

// get analog and pot value
ISR(TIMER2_OVF_vect) {
  PORTB = PORTB | 1 ;
  div2 = !div2; // reduce sampling frequency by factor 2
  if (div2) {
    selChan = !selChan;
    if (selChan) {
      potSample = ADCH;
      sbi(ADMUX, MUX0); // set muxer to ADC0
    }
    else {
      analogSample = ADCH;
      cbi(ADMUX, MUX0); // set muxer to ADC1
      adcSample = true;
    }
    sbi(ADCSRA, ADSC); // start next conversion
  }
}
```

10.6 Sound Cloud: A Digital Synthesizer

Our trip into the world of digital sound ends with a complete sound synthesizer application. It is controlled using five potentiometers.

The following circuit diagram shows the construction.

10 SOUNDS AND SYNTHESIZER

Note
The inputs of amplifiers, sound cards and powered speakers are of relatively low amplitude. Accordingly, the Arduino's output voltage must be reduced to under a volt by a voltage divider.

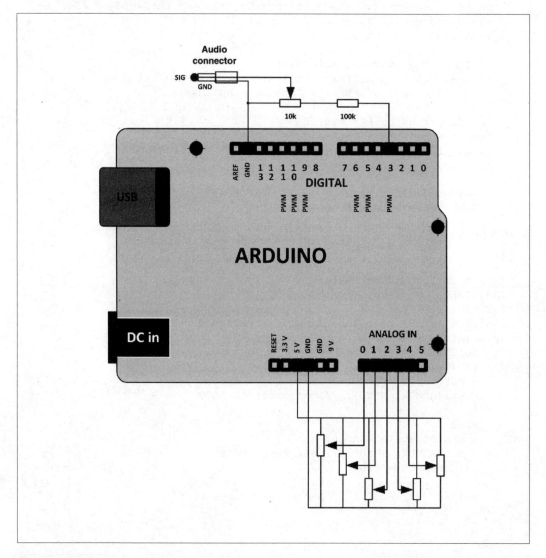

Figure 10.10: *Synthesizer*

The accompanying sketch may be downloaded from

code.google.com/p/tinkerit/downloads/detail?name=auduino_v5.pde

The sounds generated by this synthesizer are based on a broadband noise spectrum, which is repeated in rapid succession. Analog synthesizers produce similar sounds using resonant band pass filters. Noises are derived from two continuous signals, each provided with its own additional decay rate.

Five potentiometers enable the adjustment of the following parameters:

> Analog A0: Noise spectrum 1
> Analog A1: Decay rate 1
> Analog A2: Noise spectrum 2
> Analog A3: Decay rate 2
> Analog A4: Noise spectrum repeat frequency

With these five adjustment pots, an amazing range of sounds can be produced. Of course, the synthesizer is extremely extensible. By changing the basic parameters, it's possible to completely alter the sound characteristics. In addition, effects such as the reverb discussed above, as well as vibrato or various envelopes, may be created.

By using linear potentiometers instead of rotary knobs, the synthesizer can be 'played' quite well, as linear pots are easily operated using just one finger, and all five pots are easily adjusted simultaneously. A construction example is shown in Figure 10.11.

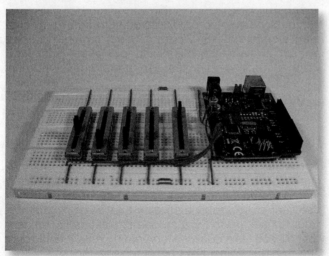

Figure 10.11:
Synthesizer Construction Example

Using Audacity, the raw sounds produced by the Arduino may be further processed, creating very interesting sound sequences for use as background ambience in videos. The appropriate backing track will certainly add a professional feel to your next YouTube video!

10 SOUNDS AND SYNTHESIZER

11
Digital Control Techniques

Basic control techniques apply not only in technology. They are found in virtually all areas of nature. Without control systems, life would almost certainly be impossible. In many biological processes, very specific conditions and values must be kept within narrow tolerances. One of the best-known control loops is the maintenance of body temperature.

Another very important feedback control loop closely controls blood sugar levels. Biological glucose sensors detect the concentration of blood glucose and the pancreas releases insulin when blood sugar levels are above normal. Insulin provides increased uptake of glucose by the body cells, and the glucose level is reduced to optimal levels. Other examples of biological control circuits are balance (walking upright), blood pressure, heart rate, pupil size, etc.

Pioneer Englishman James Watt's centrifugal governor is widely regarded as the first significant step in control technology. With it, he regulated the speed of his first steam engine. Since then, control technology has been introduced in all areas of modern life. Temperature regulators in heating systems, refrigerators and freezers, lighting automation in cameras, ABS and ESP systems in cars and even the frequency of the electrical power grid are just a few examples.

Colloquially, there is often no distinction made between control and regulation. In technology, however, these two terms have quite different meanings. Regulation refers to a process without any feedback. The output is not monitored, and may be affected by external factors. Examples may be controlling the brightness of an incandescent lamp using a dimmer, or controlling the speed of a motor using an adjustable DC voltage. In the latter case, for example, load variations may have an effect on the motor's speed, so that the intended speed is not actually achieved.

This is where control steps in. In order to keep the target speed constant, feedback is required in order to adjust the control voltage. This type of feedback is referred to as a control loop.

By 'control', a process is understood whereby an output parameter (e.g. motor speed) is monitored and an input parameter (e.g. motor control voltage) is continuously adjusted. The desired output relies on a closed circuit, or a so-called control loop.

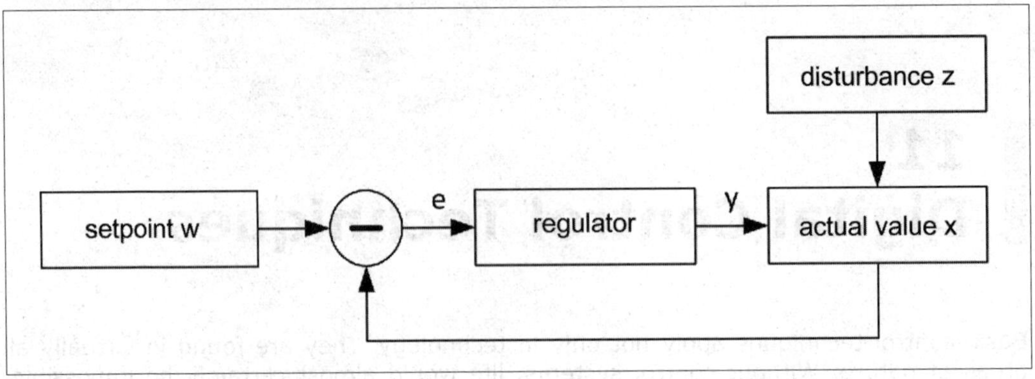

11 DIGITAL CONTROL TECHNIQUES

Figure 11.1: *Control Loop*

The key variables in a control loop are:

- the desired value, or setpoint, w
- the measured, actual value, x
- the error variable, e = w - x
- the manipulated variable, y
- the disturbance variable, z

Based on these variables, the behavior of the controller may be accurately calculated and optimized. Different applications may be achieved by combining these three basic control types.

11.1 Control Types

The task of a controller is to capture the actual value, compare it with the desired value and, by adjusting the manipulated variable in response to any deviation, keep the difference between actual and desired value as small as possible or, ideally, remove this error completely.

The timing properties and the required control precision are the most important criteria in selecting a suitable control method. According to classical control theory, a control system is made up of three basic types of controllers:

- P (proportional) controller
- I (integral) controller
- D (derivative) controller

These three types are discussed in more detail below.

11.1.1 P Controller

Pure P controllers multiply the deviation by a gain factor, Kp, and pass the result on immediately. The disadvantage of this, the simplest control type, is the final deviation that remains.

If this control type is to be software-implemented, the transfer function is

```
y = kp * error
```

11.1.2 I Controller

Integral controllers sum the deviation over a specified period of time and multiply this error (mathematically, the integral), by another factor, Ki. The result of this is that even the smallest deviations provide a significant contribution over the course of time. In ideal cases, such a deviation may ultimately be eliminated. The I controller is characterized by its high precision. Its disadvantage is that it requires some time to sum the individual deviations and react to them, so integral controllers usually have a slower response time than P controllers.

The software transfer function for an I controller is

```
errorsum = errorsum + error
y = ki * ta * errorsum
```

where `errorsum` is the sum of all previous deviations. The I controller's control parameter is dependent on the sample time, Ta. The sample time must be optimized, depending on the sampling rate.

11.1.3 PI Controller

A PI controller is a combination of P and I controllers. This type of controller combines the advantages of both, offering a comparatively fast, yet high-precision, control system.

A software PI controller would make use of the following transfer function:

```
errorsum = errorsum + error
y = kp * error + ki * ta * errorsum
```

11.1.4 PD Controller

The opposite operation to integration is differentiation. The third basic type of controller discussed was the D controller. Pure D controllers are usually quite unstable. Mathematically speaking, a D controller responds proportionally to the change in the control variable. This means that rapid changes lead to very large signals. While this enables a very fast reaction time, small or fast disturbances may lead to overcompensation, i.e. significant fluctuations in the control variable, destabilizing the control loop.

For this reason, pure D controllers are seldom used. The proportional-differential, or PD, controller is used much more often, especially when the controlled system has integral behavior characteristics. With a D controller, compensation for these characteristics is only possible to a certain extent.

11 DIGITAL CONTROL TECHNIQUES

The software transfer function for a PD controller is

```
y = kp * error + kd * (error - errorold) / ta
errorold = error
```

Here, the differentiation is approximated by a differential quotient, `(error - errorold) / ta`.

11.1.5 PID Controller

The PID controller combines all of the advantages of the fundamental controller types. This type represents the universal solution for most control problems. With optimal adjustment, the PID controller may work both very quickly and with excellent precision. Therefore, for many applications, the PID controller is the best solution.

The corresponding software algorithm for the PID controller is

```
errorsum = errorsum + error
y = kp * error + ki * ta * errorsum + kd * (error - errorold) / ta
errorold = error
```

These theoretical principles are demonstrated in the remaining sections of this chapter by means of practical examples.

11.2 Optimum Workstation Lighting: Digital Illumination Control

Our first practical example is an ambient brightness controller. There are many applications for such a controller. In many areas, uniform workplace lighting, independent of time of day, is desirable or even compulsory. The simplest form of workplace lighting control may be achieved by a two-point control. In this application, there may be, for example, sufficient natural illumination in the daytime. Electrical auxiliary lighting is turned on automatically only early in the morning or later in the evening. With a simple two-point control, certainly no absolute lighting condition is achievable.

A far better approximation may be achieved by continuous control. A light sensor, such as a photodiode, determines the ambient brightness and additional artificial light is adjusted so that a constant sum of natural and electrical lighting is maintained. Such a system is a good example for getting to know some basic controller properties.

Typical problems, such as large deviations, a slow response time or an oscillating controller are easily understood here. As Figure 11.3 shows, only a few components are required for a brightness controller. With the use of a modern high-power LED, a very efficient and practical light source may be constructed. Very usable light levels may be achieved with currents of under 500 mA. Figure 11.2 shows a small desk lamp, constructed using a 300 mA power LED.

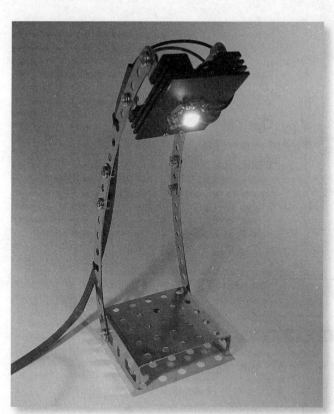

Figure 11.2:
Lamp with Power LED

The additional circuitry required for the power LED is simple, consisting of a power transistor and a resistor. A suitable light sensor may be constructed using a photodiode (e.g. BPW 40) and a resistor. An additional low-pass filter smoothes out the harmonics inherent to pulse-width modulation.

11 DIGITAL CONTROL TECHNIQUES

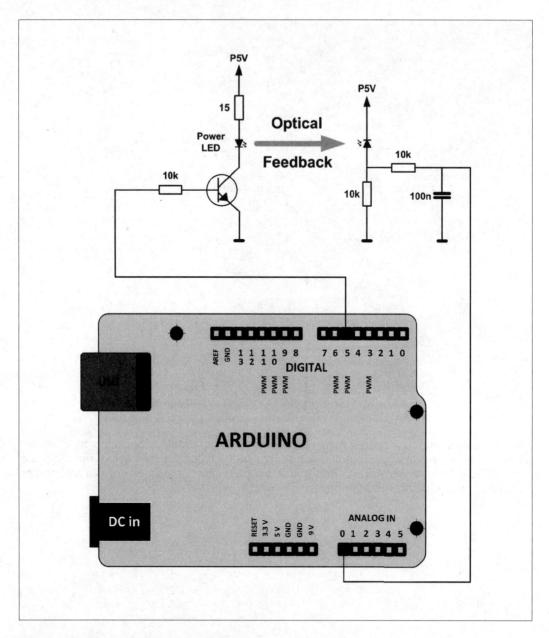

Figure 11.3: Brightness Control Circuit

Closing the control loop is simply a matter of aligning the power LED so that it illuminates the photodiode.

The circuit is designed so that the LED brightness increases with increased current, while the voltage at the analog input increases with increased light intensity. Thus, the control system is directly proportionally polarized. By calculating the error signal as

```
error = setpoint - actualValue
```

the LED brightness is throttled when the light intensity on the photodiode increases.

In the sketch, a simple proportional controller that provides very satisfactory results is implemented.

Listing 11-1: *LED Lamp Control Loop*

```
// Listing 11-1
// LED Lamp Control Loop

const int ledPin = 5;  // LED connected to digital pin 5
const int w = 100;     // set point
const int kp = 3;      // gain constant
int x, error, y;  // sensor value, error, control output

void setup() {
  pinMode(ledPin, OUTPUT);
  TCCR0B = TCCR0B & 0b11111000 | 0x01; // maximum PWM freq @ PWM5
}

void loop() {
  x = analogRead(A0);
  error = w - x;
  y = kp * error;

  if (y < 0) {
    y = 0;
  }

  if (y > 255) {
    y = 255; // range limitation
  }

  analogWrite(ledPin, y);
}
```

As starting values, a setpoint of `w = 100` and a proportional gain of `kp = 3` are chosen. By varying these two parameters, the controller can be adapted to other practical circumstances.

For this project, the PWM frequency is set to its maximum:

```
TCCR0B = TCCR0B & 0b11111000 | 0x01;
```

This sets Timer and Counter Register 0 so that the frequency divider equals 1, enabling a PWM frequency of 62.5 kHz on Channel 5.

11 DIGITAL CONTROL TECHNIQUES

Besides the control algorithm itself, the manipulated variable, y, is limited to the range of between 0 and 255, inclusive:

```
if (y < 0) y=0; if (y > 255) y=255;     // range limitation
```

This avoids excessively large or even negative values being passed to the PWM control.

11.3 A Classic of Control Theory: The Gravity Compensator

Another example of classic control theory is the so-called gravity compensator. This decorous example shows that, with the help of control techniques, the seemingly impossible may be achieved. In principle, it's not possible to get one permanent magnet to levitate in the field of another permanent magnet. Primary attractors lead to an unachievable equilibrium. While there is a point at which gravity and magnetism will exactly balance each other, the slightest deviation from this point sends the free magnet immediately toward one of the two attractors: the ground or the other fixed magnet.

One way of achieving a permanent state of levitation is to replace the fixed permanent magnet with a mounted electromagnet. By means of a programmed control loop, the electromagnetic field may be controlled such that the permanent magnet is held in suspension. This principle has many applications in engineering. A well-known example is the magnetic levitation train, which uses this technology to glide safely just a short distance above its rail.

Such control may be realized using purely analog means, but the use of digital control techniques has the advantage that the control parameters are adjusted purely in software.

Since the actuator used here is an electromagnet, the control section has an integral characteristic. As mentioned in the introductory section, the control system must have a certain 'D' component.

To determine the position of the permanent magnet, a Hall effect sensor is used. This is mounted directly below the electromagnet. At this point, the signal will certainly be influenced by both the magnetic coil and the permanent magnet. However, with a sufficiently strong permanent magnet, the influence of the electromagnetic coil is negated.

An operational amplifier is used to adjust the sensor signal. A 10 kΩ potentiometer at the non-inverting input sets the setpoint, w. The actual value, x, is fed from the Hall effect sensor to the inverting input. The op-amp generates the difference signal, e = w - x, i.e. the error signal. The gain is set using a 100 kΩ trimmer potentiometer.

The error signal, e, is sent to Arduino analog input A0. There, the manipulated variable, y, is calculated, and output to a Darlington transistor (e.g. TIP140) on Pin 5. The transistor drives the electromagnet directly. With appropriate cooling, it's capable of driving currents of several amperes. Since the Arduino is providing it with a PWM signal, the transistor's actual power dissipation remains very low, so a heat sink is unlikely to be needed.

Figure 11.4 shows the corresponding schematic diagram.

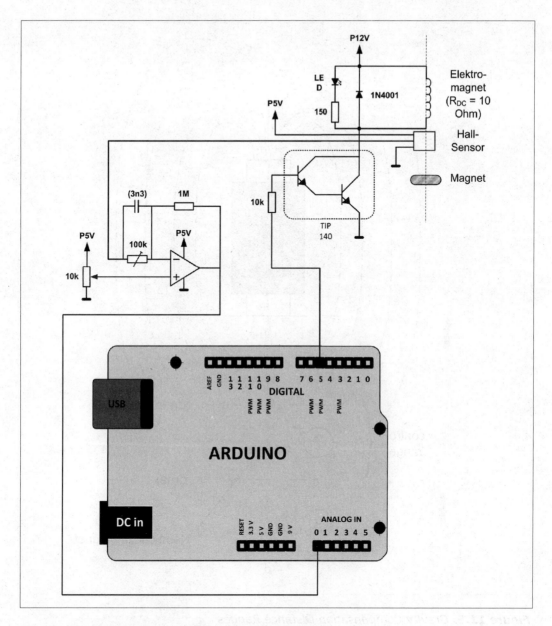

Figure 11.4: 'Gravity Compensation' Control Circuit

The typical distance between the electromagnet and the floating permanent magnet would be around 3 cm. The question surely arises: why can't this distance be increased? The reason is the electromagnet's field distribution. As Figure 11.5 shows, the field strength decreases rapidly with distance from the poles. This can only be compensated to a small extent by increasing the coil currents.

11 DIGITAL CONTROL TECHNIQUES

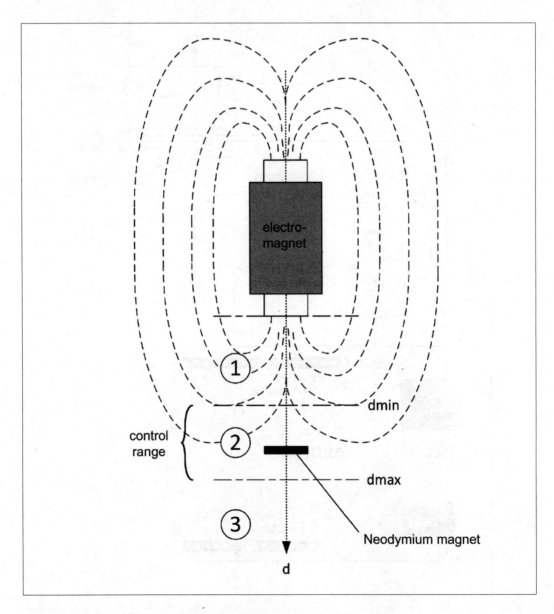

Figure 11.5: *Gravity Compensation Distance Ranges*

This experiment has three general states:

- State 1: The magnets are stuck together
- State 2: Control state
- State 3: The permanent magnet falls

A CLASSIC OF CONTROL THEORY: THE GRAVITY COMPENSATOR 11.3

As soon as system leaves State 2, it is not possible for the control system to correct it. If the system is in State 3, it cannot be corrected, regardless of how much current is supplied. If the system is in State 1, the permanent magnet becomes affixed to the electromagnet's iron core, even if the current is removed completely.

Figure 11.6 shows levitation distance as a function of coil current for a typical construction. From this, it's clear that the required coil current increases exponentially at greater distances. At distances of over 30 mm, a required current of over 2 A is quickly exceeded. This current strength carries with it a power consumption of around 40 W at a requisite coil voltage of around 20 V. This may cause significant coil heating. At even higher currents, an overheating coil may be a fire hazard.

Warning
- Do not work at very high coil currents – this is a fire hazard!
- Constantly monitor the coil temperature.
- Do not leave the gravity compensator unattended while it is switched on.

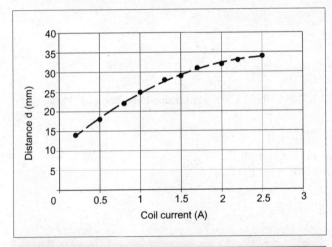

Figure 11.6:
Levitation Distance as a Function of Coil Current

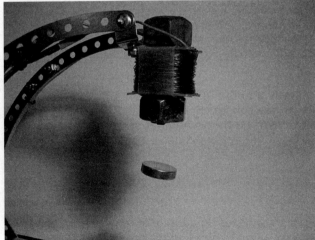

Figure 11.7:
The gravity compensator in action

11 DIGITAL CONTROL TECHNIQUES

Our sketch implements the classic PD controller. The D component compensates for the electromagnet's integral characteristic. An additional integral component arises due to the op-amp's limited bandwidth. In all, the PD control method should suffice for a stable system. The practical execution of the experiment confirms the control theory.

As in the previous section, Arduino PWM Channel 5's frequency is set at its highest, in order to achieve the maximum control speed:

```
TCCR0B = TCCR0B & 0b11111000 | 0x01;
```

Again, Timer and Counter Control Register Bit 0 is set to 1, so that a corresponding frequency of 62.5 kHz is achieved.

Listing 11-2: *Gravity Compensator*

```
// Listing 11-2
// Gravity Compensator

int error, errorOld, y;
const int kp = 1, kd = 1;

void setup() {
  pinMode(5, OUTPUT);
  TCCR0B = TCCR0B & 0b11111000 | 0x01; // maximum PWM freq @ PWM5
}

void loop() {
  error = analogRead(0) / 4;
  y = kp * error + kd * (error - errorOld);
  analogWrite(5, y);
  errorOld = error;
  // delay(10);
}
```

In the main loop, the output voltage of the Hall effect sensor is continuously measured. The sampled measurement is divided by four, so that the input range (10 bits gives us a range of between 0 and 1,023) matches the 8-bit output range (256 PWM steps).

The control signal, y, is calculated according to the PID controller formula. The initial values of `kp = 1` and `kd = 1` are set. In the construction shown in Figure 11.7, a stable state of suspension may be achieved. With other configurations, it may be necessary to adjust these values. Fine tuning of the system is done using the two potentiometers.

Tips and tricks

- The 3.3 nF capacitor in the circuit provides an additional (analog) D component. With this capacitor in use, a simple P controller may suffice on the digital side ($kd = 0$). For initial experiments, removing or inserting the capacitor may be useful in control loop optimization.

- Aerodynamic magnet stabilization:
  To find the initial levitation position, it may be useful to glue the magnet to a piece of firm paper. This increases the air resistance of the permanent magnet assembly, much like a parachute would. Precise adjustment of this slower-reacting system is often easier than with a bare magnet. Once the optimal parameters have been found, the 'paper wings' may be dispensed with.

- The LED in parallel with the coil may provide valuable assistance in adjusting the control loop. If there is no permanent magnet in the vicinity of the coil, the LED, in accordance with the coil current, lights at maximum intensity. Likewise, it dims when the permanent magnet is brought closer to the coil. As the magnet gets even closer to the coil, the LED eventually turns off. At the optimum distance, an average brightness should be obtained. The speed of the brightness variation is a good indicator of the control loop's sensitivity and may be used to aid adjustment.

When the control system is optimally adjusted, the permanent magnet engages when placed in the levitation position, as if within a tangible force field. Once this position is reached, the magnet may be released. There, it can then float freely for hours on end.

11 DIGITAL CONTROL TECHNIQUES

12
Physical Computing

The relatively new field of physical computing is concerned with collecting information about an environment and the control of mechanical parameters. While environmental data is easily collected and quantified using a variety of sensors, control of mechanical parameters is usually associated with more complexity.

The term 'physical computing' is often used in applications for art and design. The fields of robotics and autonomous machines also count as physical computing. With Arduino support, amazing projects have been realized in this field. Some of the best known are a laser harp, high-speed photography control, and a two-wheeled balancing robot.

Different systems are available for use as actuators. One of the simplest is a DC motor. At low power, it may be directly driven by an Arduino pin. At higher powers, a transistor stage is required (see Figure 5.2 for an example), as well as special suppression of the induced voltages, e.g. blocking capacitors and flyback diodes. DC motors, however, lend themselves only to relatively simple applications.

For higher precision, the use of stepper motors is recommended. These are able to be set to specific positions with a high degree of precision. However, special driver hardware is required to control stepper motors. Furthermore, by counting individual steps, only relative motion is possible. For absolute positioning, a control loop is required.

Servo motors, on the other hand, are much easier to control. These are complete driver systems that include motor control loop and positioning. In the simplest servos, the positioning system consists of a gear and a potentiometer. Because the built-in potentiometer is configured as a voltage divider and connected directly to the motor shaft via the gear, its voltage is a direct measure of the exact motor position. Thanks to its integrated control loop, a servo may be very precisely positioned with simple digital control signals. For this reason, servo motors are ideally suited to microcontroller applications.

Thus far, we have generated the usual optical and acoustic outputs using the Arduino. The following sections will expand on these into actually making physical things move and using servo motor technology to control more complex mechanical elements.

12.1 Servos Control the World

The best-known examples of servos in use are the control gear in models. In these, radio control signals are converted into control signals for the mechanics of airplane, ship or car models.

12 PHYSICAL COMPUTING

Thanks to Asian mass production, model servos are available very inexpensively. Servos are controlled by a form of pulse-width modulation, making them ideally suited for microcontroller use. Figure 12.2 shows a typical waveform.

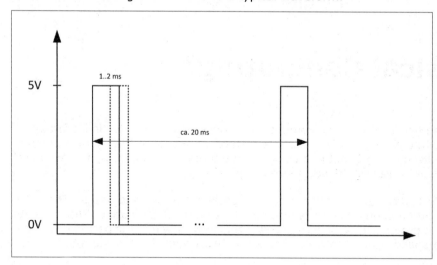

Figure 12.1: Servo Signal

The 20 ms pulse interval seen in the figure is of secondary import. Most servos can handle relatively large tolerances here. What's essential is the duration of the control pulse. A quasi-standard has emerged for this.

Pulse duration	Arm position
1 ms	Left limit
1.5ms	Center
2 ms	Right limit

Table 17: *Servo Signal Timing*

Figure 12.2: Model Servo Motors

Thanks to the already integrated control electronics, connecting a servo to the Arduino is quite easy. Other than the power supply, only a single control line is connected to one of the Arduino's PWM outputs. The individual lines are often identifiable by the following color code:

        Red: +5 V
        Black: GND
        Orange or brown: Control signal

If in doubt, consult the servo's datasheet.

Smaller servo types may be directly driven by the Arduino. For the control of larger types or even of multiple servos simultaneously, a separate 5 V source is recommended as neither the Arduino nor a USB port are guaranteed to be able to supply this much current. For this, a battery, also familiar in the modeling world, or a suitable power adapter, may be used.

Servo control is again made easy by using a library – a standard one for servos is even included with the Arduino IDE. After including the servo.h file, the function

```
xservo.write(position);
```

allows any servo to be moved to a specific position. The values for 'position' correspond to the following servo orientations:

        45:     Left extreme (-45 ° position)
        90:     Center (0 ° position)
        135:    Right extreme (+45 ° position)

In principle, larger offsets may be used, but certain servos may be damaged in that way, as the gears aren't designed to be driven past the ±45° limit.

The following sketch allows you to control a servo using a PS/2 mouse. The PS/2 interface's operation has already been described in Section 9.11, so only the servo control system functions will be described here. Once the library is included,

```
Servo xservo;
```

creates a new instance of the Servo class: an object called xservo.

Then, with

```
xservo.attach(pin);
```

this object is assigned a control output (Pin 9 in the example sketch). Finally, using

```
xservo.write(position);
```

the servo is moved.

Listing 12-1: *PS/2 Mouse-Controlled Servo*

```
// Listing 12-1
// PS/2 Mouse-Controlled Servo

#include <ps2.h>
#include <Servo.h>

// Mouse object: pin 7 is clock, 6 is data
PS2 Mouse(7, 6);

const int minPos = 45, midPos = 90;
char mouseStat, mouseX, mouseY; // mouse status
Servo xServo;         // create x servo object
int xPos;             // servo position
int xOffset = -20;  // offsets

void setup() {
  Serial.begin(9600);
  mouseInit();
  xServo.attach(9);    // x servo on pin 9
}

void loop() {
  // get mouse info
  Mouse.write(0xeb);     // request data
  Mouse.read();  // ignore confirm
  mouseStat = Mouse.read();
  mouseX = Mouse.read();
  mouseY = Mouse.read();

  // move x servo
  xPos = xPos + mouseX;

  if (minPos <= xPos) {
    xPos = minPos;
  }
  if (minPos >= xPos) {
    xPos = -minPos;
  }

  Serial.print(" xPos: ");
  Serial.println(xPos);
  xServo.write(xOffset + midPos + xPos);
}

void mouseInit() {
  Mouse.write(0xff);            // reset
  Mouse.read();                 // confirm byte 3x
  Mouse.read(); Mouse.read();
  Mouse.write(0xf0);            // remote mode
```

```
    Mouse.read();
    delayMicroseconds(100);
}
```

12.2 'Photino': 2D Camera Swivel

For our first useful physical computing application, two servos are required. These are mounted at right angles to each other. When using X-shaped servo horns, a reasonably secure attachment is achievable using sturdy rubber bands. For the rest of the mechanism, strips of wood may be used. To allow scope for further modification and optimization, simple toy mechanical construction systems may be used. Kits with perforated plates and matching screws are available at a low cost. These may be used in other projects to extend electronics using mechanical systems.

Figure 12.3 shows a 2D camera positioner made with the help of mechanical constructor system components. The camera is attached to an elongated servo horn.

In the sketch, the following two required libraries are used:

```
ps2.h
Servo.h
```

After specifying the mouse pin connections, the mouse is initialized. The servo port pins are specified and the servos are initialized.

In the main loop, mouse movements are recorded in both the X and the Y directions. From these, the servo positions are calculated. In order not to drive the servos past their mechanical limits, the position values are software-limited to the ±45° range.

After sending the values to the serial interface for debugging purposes, the servo positions are calculated in relation to the center and offset values.

With this structure, a compact camera may be very accurately pointed in two dimensions. All sorts of surveillance tasks are potential applications. If a good USB cable is used, several meters may be spanned, so control of a camera in the next room poses little problem. Using a webcam, an entire room may be monitored by turning the camera head.

This application will allow you to keep an eye on small children in the next room, or ensure that the beloved house pet is not destroying the living room while you're in the study.

A camera-panning device is also ideal for panoramic photographs. With a suitably sturdy construction, wide mountain vistas or large cloud structures may be scanned.

12 PHYSICAL COMPUTING

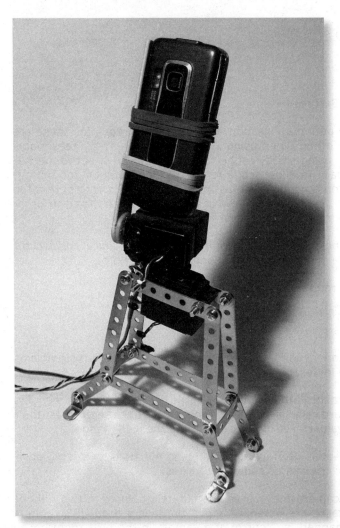

Figure 12.3:
Photino

Listing 12-2: Photino

```
// Listing 12-2
// Photino

#include <ps2.h>
#include <Servo.h>

const int xOffset = -20, yOffset = -20;  // offsets
const int midPos = 90, maxPos = 135;
const int delTm = 300;                    // delay time
Servo xServo;                             // create x-servo object
Servo yServo;                             // create y-servo object
```

226

'PHOTINO': 2D CAMERA SWIVEL 12.2

```
int xPos, yPos;                       // servo position
char mouseStatus, mouseX, mouseY;     // mouse status

// Mouse object: 7 is clock, 6 is data
PS2 Mouse(7, 6);

void setup() {
  Serial.begin(9600);
  mouseInit();
  xServo.attach(9);                   // x-servo on pin 9
  yServo.attach(10);                  // y-servo on pin 10
  pinMode(13, OUTPUT);                // magnet control
}

void loop() {
  // get mouse info
  Mouse.write(0xeb);                  // request data
  Mouse.read();                       // ignore ack
  mouseStatus = Mouse.read();
  mouseX = Mouse.read();
  mouseY = Mouse.read();

  // move x servo
  xPos=xPos+mouseX;
  if (xPos>=45) xPos=45;
  if (xPos<=-45) xPos=-45;
  Serial.print(" xPos: ");
  Serial.print(xPos);
  xServo.write (xOffset+midPos+xPos);

  // move y servo
  yPos=yPos+mouseY;
  if (yPos>=45) yPos=45;
  if (yPos<=-45) yPos=-45;
  Serial.print(" yPos: ");
  Serial.print(yPos);
  yServo.write(yOffset+midPos+yPos);
}

void mouseInit() {
  Mouse.write(0xff);                  // reset
  // conf byte 3x
  Mouse.read(); Mouse.read(); Mouse.read();
  Mouse.write(0xf0);                  // remote mode
  Mouse.read();
  delayMicroseconds(100);
}
```

12.3 'Cranino': Mouse-Controlled Crane

Another application that will amuse children or the young at heart is the 'Cranino'. The mechanical construction is virtually identical to that of the Photino, except that a load arm with an attached electromagnet is attached to the servo instead of a camera.

The electromagnet is then driven using one of the other Arduino pins. Since relatively large currents are used, the by-now-familiar transistor stage is employed.

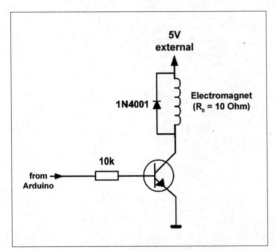

Figure 12.4:
Transistor Stage for the Electromagnet

Power for the electromagnet should never be supplied by the Arduino board. A typical electromagnet has a resistance of only 10 Ω, so, at 5 V, it consumes 500 mA of current. This exceeds what's available on most USB ports, as well as the Arduino's voltage regulator if an external power supply is used to power the Arduino. For this reason, it's necessary to supply the electromagnet with its own power from a small external power supply, or a battery if need be.

The magnet is controlled using the left mouse button. The Cranino sketch is essentially the same as the Photino one, with control of the electromagnet added. Both the left mouse button's status is queried and the electromagnet activated or deactivated by the code

```
digitalWrite(13, (mouseStatus & 0b0001));
```

With a little practice, the Cranino can be used to move paper clips or other metal objects from one place to another.

'CRANINO': MOUSE-CONTROLLED CRANE 12.3

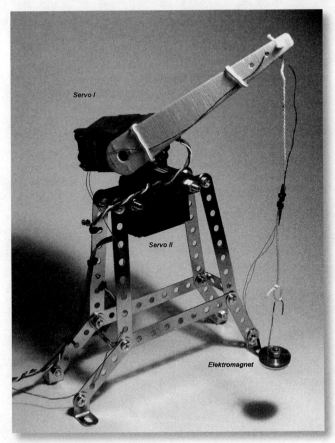

Figure 12.5:
Cranino

Listing 12-3: Cranino

```
// Listing 12-3
// Cranino
#include <ps2.h>
#include <Servo.h>

// Mouse object: pin 7 is clock, 6 is data
PS2 Mouse(7, 6);

Servo XServo;              // create x-servo object
Servo YServo;              // create y-servo object
int xPos = 0, yPos = 0; // servo position
int xOffset = -20, yOffset = -20; // offsets
int minPos = 45, midPos = 90, maxPos = 135;
int delTm = 300;           // delay time
char mouseStatus, mouseX, mouseY; // mouse status

void setup() {
  Serial.begin(9600);
```

```
  mouseInit();
  XServo.attach(9);     // x-servo on pin 9
  YServo.attach(10);    // y-servo on pin 10
  pinMode(13, OUTPUT);  // magnet control
}

void loop() {
  // switch magnet on Pin 13 if left mouse button is pressed
  digitalWrite(13, (mouseStatus&0b0001));

  // get mouse info
  Mouse.write(0xeb);    // request data
  Mouse.read();         // ignore ack
  mouseStatus = Mouse.read();
  mouseX = Mouse.read();
  mouseY = Mouse.read();

  // move x servo
  xPos = xPos + mouseX;
  if (xPos >= 45) {
    xPos = 45;
  }

  if (xPos<=-45) {
    xPos = -45;
  }

  Serial.print(" xPos: ");
  Serial.print(xPos);
  XServo.write(xOffset + midPos + xPos);

  // move y-servo
  yPos = yPos + mouseY;
  if (yPos >= 45) yPos=45;
  if (yPos <= -45) yPos=-45;
  Serial.print(" yPos: ");
  Serial.print(yPos);
  YServo.write(yOffset + midPos + yPos);
}

void mouseInit() {
  Mouse.write(0xff);    // reset
  // ack byte 3x
  Mouse.read(); Mouse.read(); Mouse.read();
  Mouse.write(0xf0);    // remote mode
  Mouse.read(); delayMicroseconds(100);
}
```

13
Processing

Processing is the predecessor (and PC programming counterpart) to Arduino IDE. Owing to its legacy, you may sometimes hear the Arduino environment itself being referred to as Processing. In Figure 13.1, the similarity between the two applications is clear.

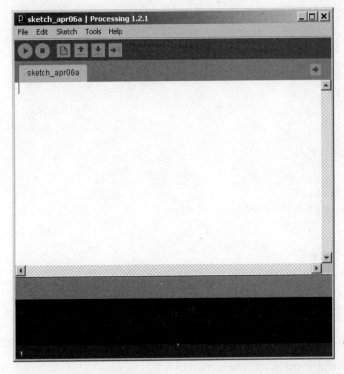

Figure 13.1:
The Processing Interface

The program's icons have the same meanings as with Arduino IDE, with the difference being that the 'Run' icon doesn't upload the sketch to an Arduino. Rather, the sketch is run directly on the PC.

Processing is free, open source and available for download from

processing.org/

As with the Arduino IDE, many interesting example programs are included. Just trying

13 PROCESSING

out all of these examples is already an extensive, but very interesting and instructive task.

13.1 Arduino and Processing: A Formidable Team

Processing was developed to give students who were not in the information science field the ability to create complex and impressive animation and other graphics. The included example sketches offer a fascinating first impression.

Processing is also ideal for use in conjunction with the Arduino. For an Arduino user, working with Processing is even easier, since many of the basics are already known.

Processing sketches also require two basic functions: `void setup()`, and `void draw()` instead of `void loop()`.

A simple 'Hello World' sketch looks like this:

Listing 13-1: *Processing Hello World (Processing .pde)*

```
// Listing 13-1
// Processing Hello World

int x = 30;
PFont FontA;

void setup() {
  size(200, 120);
  background(102, 100, 210);
  FontA = loadFont("Ziggurat-HTF-Black-32.vlw");
  textFont(FontA, 32);
  noLoop();
}

void draw() {
  fill(200, 100, 0); text("Hello", x, 60);
  fill(250, 10, 30); text("World", x, 95);
}
```

This sketch creates a rather colorful window on the screen.

With the right libraries, Arduino and Processing make up a formidable team. Detailing the Processing language in detail is beyond the scope of this Chapter. Rather, in the sections that follow, two example applications will be explained that demonstrate how Arduino and Processing can work together to create graphical representations of data.

From these foundations, the applications may easily be incrementally extended.

SIMPLE BEGINNINGS: THE WARNING LIGHT 3.3

Note
Like the Arduino versions prior to 1.0, sketches for Processing have the `.pde` extension. This may cause some confusion, but it helps somewhat that the Arduino IDE has switched to the `.ino` extension since Version 1.0.

13.2 Interaction with Processing: Data Logging, Trend Graphs, etc.

In the first project, all six of the Arduino's analog channels are be displayed simultaneously on the PC monitor in the form of a bar graph. A screenshot is shown in Figure 13.2.

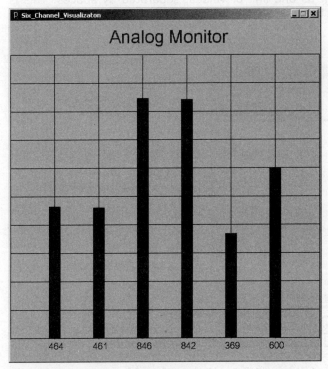

Figure 13.2: Analog Monitoring

The Arduino sketch is very simple.

Listing 13-2: 6-Channel Analog Monitor (Arduino .ino)

```
// Listing 13-2
// 6-Channel Analog Monitor
// ARDUINO Sketch

void setup() {
  Serial.begin(115200);
}

void loop() {
  // build message with values of analog inputs (0-5)
```

```
  for(int i = 0; i < 6; i++) {
    Serial.print(analogRead(i));
    Serial.print(" ");
  }
  Serial.println();
}
```

First, the serial interface is initialized to 115,200 baud. In the main loop, the analog-to-digital converter channels are read and the values sent to the serial interface. The six measured values are sent as one line of values, separated by spaces. This is followed by a line break and then the next six values, etc. You can see the result of this in Serial Monitor.

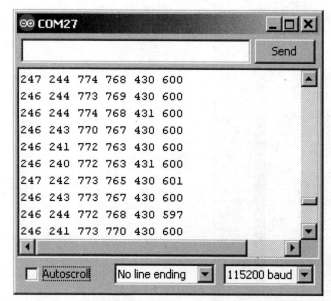

Figure 13.3:
Output of Analog Data to Serial Monitor

On the PC side, the Processing sketch is a little more complex. After some basic definitions, such as the width and height of the output window, the `processing.serial` library is imported. This makes it possible to communicate with the Arduino.

The two functions, `drawGrid()` and `serialEvent()` follow. The first of these creates a coordinate graph and writes the title to the active window. `serial_event()` gets the analog values from the serial port and saves them in an array of 6 elements.

In `setup()`, the active window is created and communication with the serial interface is begun.

The main function, `draw()`, then draws the values as six bars. The height of each bar is proportional to the input values at each ADC.

Listing 13-3: *6-Channel Analog Monitor (Processing .pde)*

```
// Listing 13-3
// 6-Channel Visualization
// PROCESSING Sketch

// basic data
int xMax = 561;
int yMax = 600;
int offset = 60;
int distX = 80;
int distBot = 40;
int colWidth = 20;
PFont FontA;
int[] sensorValues = new int[6];

// libraries
import processing.serial.*;

// serial connection
Serial Port;
String message = null;
String elements[] = null;
color graphColor = color(25, 25, 250);
PFont fontGraph;

// draw grid and title
void drawGrid() {
  background(200);
  stroke(0);

  // vertical line
  for (int x = 0; x <= xMax; x += distX) {
    line (x,offset,x,xMax);
  }

  // horizontal lines
  for (int y = 0; y <= yMax; y += 50) {
    line (0, y + offset, yMax, y + offset);
  }
  textFont(FontA, 32);
  fill(graphColor);
  text("Analog Monitor", xMax / 2 - 100, 40);
  textFont(FontA, 16);
}

void serialEvent(Serial p) {
  message = Port.readStringUntil(13);
  if (message != null) {
    try {
      elements = splitTokens(message);
```

```
          for (int i = 0; i < elements.length && i < 6; i++) {
            sensorValues[i] = int(elements[i]);
          }
        }
        catch (Exception e){}
      }
  }

  void setup() {
    size(xMax, yMax);
    noStroke();
    fontGraph = loadFont("ArialUnicodeMS-48.vlw");
    textFont(fontGraph, 12);
    // show values
    println(Serial.list());
    Port = new Serial(this, "COM5", 115200);
    FontA = loadFont("ArialUnicodeMS-48.vlw");
    textFont(FontA, 16);
  }

  void draw() {
    drawGrid();
    for (int i = 0; i < 6; i++) {
      fill(graphColor);
      rect(i * distX + distX - colWidth / 2, height - distBot, colWidth,
  -sensorValues[i] / 2);
      text(sensorValues[i], i * distX + distX - colWidth / 2, height -
  distBot + 20);
    }
  }
```

To test the program, six potentiometers (e.g. 10 kΩ) may be hooked up to the Arduino. In this example, we have used PC serial port COM3. Remember to change this to the port to which your Arduino is connected.

Extensions and exercises

If the potentiometers are replaced with temperature sensors (see Section 7.7), the temperatures in six different rooms may be monitored simultaneously.

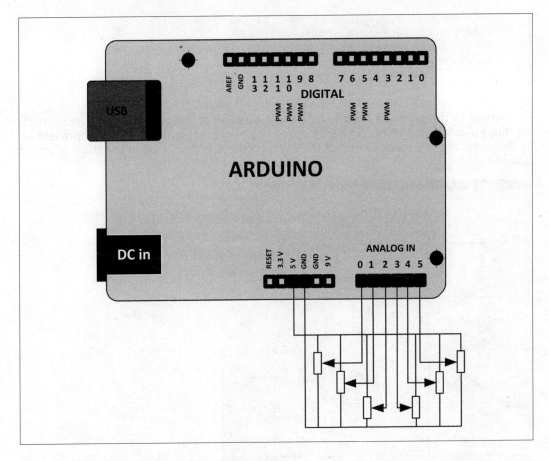

Figure 13.4: *Test Rig for All 6 Analog Channels*

In the second project, only one analog channel is used. This will be graphed over a course of time. Systems that capture measured values over specific time periods and enable them to be viewed later are known as data loggers.

Listing 13-4: *Data Logger (Arduino .ino)*

```
// Listing 13-4
// Data Logger

int pinPot, valPot;

void setup() {
  // initialize serial transmission
  Serial.begin(115200);
}

void loop() {
```

13 PROCESSING

```
    // send analog readings to PC - range: 0-255
    valPot = analogRead(pinPot) >> 1;
    Serial.println(valPot);
    delay(50);
}
```

In contrast to the previous sketch, here the X position of the current data point is shifted three pixels to the right. No bar is drawn; rather, a single data point is represented. In this way, the temporal course of the measurement is displayed in the form of an x-t diagram.

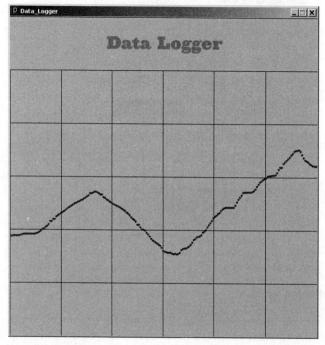

Figure 13.5:
Data Logger

Listing 13-5: Data Logger (Processing .pde)

```
// Listing 13-5
// Data Logger

import processing.serial.*;

// basic data
int xMax = 600; int yMax = 600;
PFont FontA;

// read value from Arduino
int valPot;
float valFloat;
int xPos = 0;
```

238

```
// ASCII linefeed character (end of string)
int linefeed = 10;
Serial Serport;

// draw grid and title
void drawGrid() {
  background(200); stroke(0);
  // vertical lines
  for (int x = 0; x <= xMax; x += 100) {
    line (x, 100, x, xMax);
  }

  // horizontal lines
  for (int y = 0; y <= yMax; y += 100) {
    line (0, y, yMax, y);
  }

  fill(250, 100, 0);
  text("Data Logger", xMax / 2 - 110, 60);
}

void setup() {
  // set window size
  size(xMax, yMax);
  Serport = new Serial(this, "COM3", 115200);
  FontA = loadFont("Ziggurat-HTF-Black-32.vlw");
  textFont(FontA, 32);
  background(200);
  drawGrid();
}

void draw() {
  // check data
  while (Serport.available() > 0) {
    String valString = Serport.readStringUntil(linefeed);
    if (valString != null) {
      print(valString);
      valFloat = float(valString);
      valPot = int(valFloat);
    }
  }
  fill(255, 25, 0);
  ellipse(xPos, yMax - valPot -3, 3, 3); // set circle dot
  xPos = xPos + 3;
  if (xPos > xMax) {
    xPos =0;
    drawGrid();
  }
  Serport.clear();
  delay(30);
}
```

13 PROCESSING

14
The 'Living Room Box': Our Modular Concluding Project

In the final chapter, a somewhat larger project will be presented. We call it the 'Living Room Box' – a fully practical device that may be modified to the owner's needs and utilized for new applications.

The box's modular design makes it flexible enough to suit it to the owner's needs. Of course, the features presented here are just a tiny sampling of the possible applications. More sensors, displays, etc., may be added, depending on your needs. With the knowledge acquired from reading this book, this should present no problems.

With a neat 32 cm x 22 cm x 8 cm enclosure, this attractive box should attract no objections from one's better half about keeping home-made electronics in the living room. All of the information provided by the box may be displayed on the LCD. For extensibility and upgradability, a 4x16- or 4x20-character display is recommended.

To control a variety of applications with a single controller, the I²C bus is used extensively.

Figure 14.1:
The Living Room Box

In addition to the Arduino, the box is able to house a large breadboard. This serves as the base for all of the peripherals. It's helpful if the board is of the type with additional power rails above and below the usual, main breadboarding area. Using these rails, the power supply and SDA and SCL signals are easily distributed. These rails are usually attached to the main board, and are removable.

14 THE 'LIVING ROOM BOX': OUR MODULAR CONCLUDING PROJECT

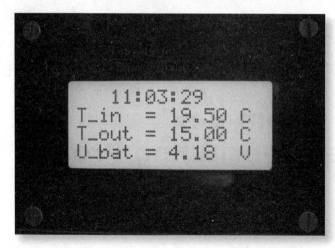

Figure 14.2:
Display Showing Time, Indoor and Outdoor Temperature, as well as Battery Voltage.

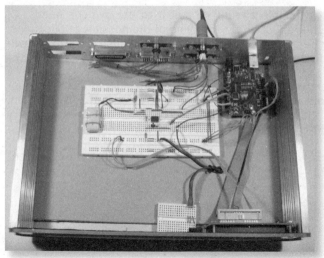

Figure 14.3:
Living Room Box Internals

The Living Room Box's hardware is described in Section 14.7.

14.1 Always Useful: A Clock

The basis of the living room box is a clock. A timekeeper should not be absent in any living room, lest one miss one's favorite TV shows. The time is displayed on the LCD in hours, minutes and seconds. Improved clock accuracy is achieved by using a watch crystal.

A PCF8583 real-time clock chip serves as the heart of the clock. It is backed up by a 3.8 V NiMH battery. The relevant details are in Section 9.7. When the Arduino is provided with external power, the battery is automatically charged. Upon power failure, say, if the box is unplugged to be moved to another location, the clock continues to run, and, after reconnecting the supply voltage, the correct time is still displayed.

Extensions and exercises

- How might the clock's accuracy be further improved?
- Upgrade the clock to make use of a DCF77 or GPS module for time. The real-time clock may still be used as a backup, in cases where the signal can't be received with sufficient quality due to poor reception conditions, etc.

14.2 Control from Afar: The IR Interface

We equipped the Living Room Box with an infrared remote control so that it doesn't have to always be within arm's reach. The details of this interface have been explained in Section 9.8. In our example sketch, the clock is set using the IR interface. The following keys are used to set the time:

1:	Hours+
4:	Hours-
2:	Minutes+
5:	Minutes-

The IR receiver can be placed directly behind a transparent plastic screen, although it's wise to ensure that the screen is also transparent to infrared light, in order that the operating range not be reduced. If the screen isn't suitable for infrared reception, an additional recess may be needed in the enclosure specifically for the IR receiver.

Extensions and exercises

Remote control is obviously suitable for other functions:

- Could you set it up to control the LC display's backlighting?
- If the display is not large enough to display all of the available / relevant information, the information of interest may be selected using a small menu.

14.3 230 V Control for Hi-Fi systems, Televisions, Lamps, etc.

Since the box contains an accurate clock, control of mains-powered devices would be an interesting option. As demonstrated in Section 8.5, mains-powered appliances are easily and safely controlled using a switching device. On the sketch side, we simply compare the current time with the required switching time, which may, of course, be set with the IR remote. Turning on ordinary lamps, or a radio or television with second accuracy is easy. For example, we might ensure that a specific lamp is turned on every day at 21:00, or that the TV turns off every night at 23:00. All kinds of alarms and timers may be implemented as well.

14.4 Timers and Sensors as the Basis for Home Automation

Still more impressive would be complete home automation. Progress in this direction has been quite successful of late. A timer is not the only way of controlling electrical appliances. Using a light sensor, for example, a lamp may be turned on only when the room gets too dark. Control of electrical heaters depending on the room temperature, or automatically opening and closing windows at specific room humidity or temperatures are other examples.

The last two sections build some foundations for this convenience. An integrated temperature and humidity sensor is presented as an example.

14.5 Indoor and Outdoor Thermometers

Besides the time, a temperature display is a classic application for this living room box. To reduce the number of Arduino pins required, use is made of an I²C temperature sensor. As shown in Section 9.6, the LM75 sensor is ideal for this task. In the example program, two of these sensors are used, allowing indoor and outdoor temperatures to be measured simultaneously. One sensor is mounted within (or close to) the box for the indoor measurement, and the other may be connected via a long cable.

Additional sensors are easily added — in a larger house, the temperature in different rooms, the cellar or the attic, may be measured and then read from a central location. When certain thresholds are exceeded, different alarms may be triggered. For example, if the wine cellar gets cold enough for the wine to spoil, or if the storage attic's temperature exceeds 50 °C, the alarm could prompt you to move your delicates and perishables to a more suitable location in the house.

14.6 No More Dry Air: A Hygrometer

If humidity is monitored alongside temperature, it's easier to ensure optimal room climate. In basements, for example, excessive humidity may lead to mold growth. Putting a humidity sensor along with a temperature sensor there provides us with has some assistance in preventing mold growth.

The operation and assessment of the SHSA2 humidity sensor was discussed in Section 7.11. By combining temperature and humidity readings, a 'mold alarm' may be triggered, calling for the opening of the cellar window at certain temperature / humidity conditions.

Humidity also plays a crucial role in the maintenance of a healthy indoor climate. For certain people, such as people with allergies, newborn babies, or the elderly and frail, room climate should be closely monitored. With the Living Room Box and suitable sensors, this presents no problem at all.

14.7 The Hardware

All of the hardware used in this project has been covered in previous sections in this book. The following figure shows a block diagram that includes some optional extras.

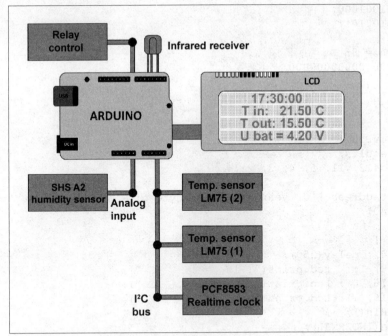

Figure 14.4:
Living Room Box Block Diagram

14.8 Living Room Box Example Program

Despite its extensive functionality, the sketch for the living room box is actually quite compact, thanks to the use of libraries. In the sketch, after including the libraries, a template is created in `setup()` for the information to be displayed on the LCD.

In the main loop, the current time is displayed first. The rest of the sensors are then queried, and the results written to the display. The real-time clock's battery voltage is also checked and displayed.

As mentioned earlier, the time is conveniently adjustable using an infrared remote control. This means the Living Room Box can get by without any of the classic controls such as knobs or dials.

Listing 14-1: *Living Room Box*

```
// Listing 14-1
// Living Room Box

#include <Wire.h>
#include <stdio.h>
#include <IRremote.h>
#include <PCF8583.h>
#include <LiquidCrystal.h>
```

```
#define LM75_1 0b1001000
#define LM75_2 0b1001011

const int recvPin = 8;
const long tB = 0b100000000000;

float vBat;
int irCode;

IRrecv IrRecv(recvPin);
decode_results results;
LiquidCrystal Lcd(12, 11, 5, 4, 3, 2);

const int correctAddress = 0, yearOf = 2000; byte of = 48, mu = 10;
PCF8583 Clock(0xA0);

void setup(void) {
  Lcd.begin(16, 4); delay(250);
  Lcd.setCursor(0, 1);  Lcd.print("T_in  =      C ");
  Lcd.setCursor(16, 0); Lcd.print("T_out =      C ");
  Lcd.setCursor(16, 1); Lcd.print("V_bat =      V ");
  IrRecv.enableIRIn();
  analogReference(INTERNAL);
}

void loop(void) {
  Wire.begin();
  Clock.get_time();
  char time[10];
  sprintf(time, "%02d:%02d:%02d", Clock.hour, Clock.minute, Clock.second);
  Lcd.setCursor(3, 0); Lcd.print(time);

  byte msb1, lsb1 = 0; byte msb2, lsb2 = 0;
  float degrees1 = 0;  float degrees2 = 0;

  Wire.beginTransmission(LM75_1);
  Wire.send(0x00);
  Wire.endTransmission();
  Wire.requestFrom(LM75_1, 2);
  while(2 > Wire.available());
  msb1 = Wire.receive();
  lsb1 = Wire.receive();
  if (0x80 > msb1) {
    degrees1 = ((msb1 * 10) + (((lsb1 & 0x80) >> 7) * 5));
  }
  else {
    degrees1 = ((msb1 * 10) + (((lsb1 & 0x80) >> 7) * 5));
    degrees1 = -(2555.0-degrees1);
  }
  degrees1 = degrees1 / 10;
```

```
  Wire.beginTransmission(LM75_2);
  Wire.send(0x00); Wire.endTransmission();
  Wire.requestFrom(LM75_2, 2); while(2 > Wire.available());
  msb2 = Wire.receive();
  lsb2 = Wire.receive();
  if (msb2 < 0x80) {
    degrees2 = ((msb2 * 10) + (((lsb2 & 0x80) >> 7) * 5));
  }
  else
  {
    degrees2 = ((msb2 * 10) + (((lsb2 & 0x80) >> 7) * 5));
    degrees2 = -(2555.0-degrees2);
  }
  degrees2 = degrees2 / 10;

  Lcd.setCursor(8, 1);
  Lcd.print(degrees1);
  Lcd.setCursor(4, 2);
  Lcd.print(degrees2);
  delay(100);

  vBat = analogRead(0) * 12.1 / 1023;
  Lcd.setCursor(4, 3);
  Lcd.print(vBat);

  if (IrRecv.decode(&results)) {
    irCode = results.value;
    if (irCode >= tB) {
      irCode -= tB;
    }

    switch (irCode) {
      case 1:
        if (23 > Clock.hour) { Clock.hour++; } else { Clock.hour = 0; }
        break;
      case 4:
        if (0 < Clock.hour) { Clock.hour--; } else { Clock.hour = 23; }
        break;
      case 2:
        if (59 > Clock.minute) { Clock.minute++; } else { Clock.minute = 0; }
        break;
      case 5:
        if (0 < Clock.minute) { Clock.minute--; } else { Clock.minute = 59; }
        break;
    }
```

14 THE 'LIVING ROOM BOX': OUR MODULAR CONCLUDING PROJECT

```
        IrRecv.resume();
        Clock.set_time();
    }
}
```

Bibliography

[1] Brian W. Kernighan, Dennis M. Ritchie; The C Programming Language, Prentice-Hall Int., (1990).

[2] P. Horowitz, W. Hill; The Art of Electronics, Cambridge University Press (1989).

[3] U. Tietze, C. Schenk; Electronic Circuits, Springer-Verlag (1991).

Supplier Directory

Electronic components may be obtained from the following companies:

- Mouser: www.mouser.com
- DigiKey: www.digikey.com
- RS Components: www.rscomponents.com
- Element14 / Farmell /Newark: www.element14.com

Sketches, Information and Updates

The source code for all of the sketches contained herein are available for download from the Elektor website, at:

    elektor.com/ArduinoProjects

The sketch file names begin with a reference to the listing number in the book, for example:

    Listing_0501_LED_Chaser.ino

At the website, further information and updates are published, as far as possible. If a version on the website differs from that printed in the book, one may assume that the online version has been updated and contains the correct code.

Listings

Listing 5-1:	LED Chaser	52
Listing 5-2:	Persistence of Vision	56
Listing 6-1:	Multiplexed 7-Segment Display	61
Listing 6-2:	Test 4x7 LED Display	64
Listing 6-3:	LedDisplay.h	65
Listing 6-4:	Dot Matrix Graphic Display	67
Listing 6-5:	Dot Matrix Counter Display	71
Listing 6-6:	Dot Matrix Character Display	73
Listing 6-7:	LCD Display Test	79
Listing 7-1:	Bargraph Voltmeter	82
Listing 7-2:	LCD Voltmeter	85
Listing 7-3:	Kiloohmmeter	89
Listing 7-4:	Elcaduino	90
Listing 7-5:	Picofaraduino	92
Listing 7-6:	Transistino	94
Listing 7-7:	NTC Thermometer	97
Listing 7-8:	AD22100 Thermometer	99
Listing 7-9:	Thermometer with 7-Segment Display	102
Listing 7-10:	LedDisplayDegrees.h	102
Listing 7-11:	SHS A2 Hygrometer	106
Listing 7-12:	Battduino	109
Listing 7-13:	Reflex Flasher	114
Listing 7-14:	Digital Luxmeter	117
Listing 7-15:	Ultrasonic Radar	120
Listing 8-1:	Grand Prix Teeth-brushing Timer	126
Listing 8-2:	Adjustable 7-Segment Display Clock	132
Listing 8-3:	LedDisplayV3.h	133
Listing 8-4:	Reaction Timer	137
Listing 8-5:	Timer with 4x7-Segment Display	140
Listing 8-6:	DCF77 Decoder with Serial Output	146
Listing 8-7:	DCF77 Clock with LCD Display	147
Listing 9-1:	PCF8574 Mega LED Chaser	153
Listing 9-2:	PCF8574 Hexadecimal Debugger Display	156
Listing 9-3:	I2C LCD Test	158
Listing 9-4:	LM75 LCD Thermometer	162
Listing 9-5:	Real-Time Clock with LCD	165
Listing 9-6:	IR Receiver	170

Listing 9-7:	IR-Controlled RGB Lamp	173
Listing 9-8:	LED Clock with Infrared Remote	176
Listing 9-9:	PS/2 Mouse Control	180
Listing 9-10:	PS/2 Keyboard and LCD	182
Listing 10-1:	Simple Alarm Sound	185
Listing 10-2:	Red Alert	186
Listing 10-3:	Fast PWM Sine Wave	189
Listing 10-4:	Fast PWM Waveforms	191
Listing 10-5:	Fast PWM Bell Sound	193
Listing 10-6:	Theremin	197
Listing 10-7:	Voltage-Controlled Oscillator	199
Listing 10-8:	Reverb Generator	202
Listing 11-1:	LED Lamp Control Loop	213
Listing 11-2:	Gravity Compensator	218
Listing 12-1:	PS/2 Mouse-Controlled Servo	224
Listing 12-2:	Photino	226
Listing 12-3:	Cranino	229
Listing 13-1:	Processing Hello World (Processing .pde)	232
Listing 13-2:	6-Channel Analog Monitor (Arduino .ino)	233
Listing 13-3:	6-Channel Analog Monitor (Processing .pde)	235
Listing 13-4:	Data Logger (Arduino .ino)	237
Listing 13-5:	Data Logger (Processing .pde)	238
Listing 14-1:	Living Room Box	245

List of Figures

Figure 2.1:	Arduino Uno SMD	11
Figure 2.2:	Arduino Pinout	13
Figure 2.3:	Typical DIY shield	14
Figure 2.4:	Power Supply Connector Polarity	15
Figure 3.1:	Typical Arduino Directory	18
Figure 3.2:	Arduino Shortcut for Starting the IDE	18
Figure 3.3:	Arduino IDE 1.0.5 Splash Screen	19
Figure 3.4:	Empty Sketch Window	19
Figure 3.5:	Selecting the Correct Arduino Version	20
Figure 3.6:	Selection of the Virtual COM Port	20
Figure 3.7:	IDE Icons	21
Figure 3.8:	Blink Sketch Loaded in the IDE	22
Figure 3.9:	LED 13 in action	23
Figure 3.10:	Arduino IDE on Ubuntu	23
Figure 3.11:	Selection of the Serial Interface under Linux	24
Figure 3.12:	Online Help	28
Figure 3.13:	Excerpt of the Help Text for the digitalWrite() Function	28
Figure 4.1:	Breadboard	44
Figure 4.2:	Stripboard	44
Figure 4.3:	A Perfboard has Individual Pads	44
Figure 4.4:	Stripboard Circuit	45
Figure 5.1:	LED Chaser	52
Figure 5.2:	Controlling a Power LED	54
Figure 5.3:	Circuit Diagram for POV display	55
Figure 5.4:	Construction of the POV Display	55
Figure 5.5:	POV 'HELP' Display	55
Figure 6.1:	Typical Pin Layout on a Single 7-Segment Display	60
Figure 6.2:	Control of a Single 7-Segment Display.	60
Figure 6.3:	7-Segment Display in Action	61
Figure 6.4:	Integrated 4-Digit, 7-Segment Numeric Display	63
Figure 6.5:	4-Digit, 7-Segment Common Anode Display Connected to Arduino	64
Figure 6.6:	Mini Graphical Display Circuit	70
Figure 6.7:	Dot Matrix Display as a Two-Digit Digital Display	71
Figure 6.8:	Dot Matrix Display of an Alphabetical Character	73
Figure 6.9:	4-Line, 16-Character LCD Connected to the Arduino	78
Figure 6.10:	LCD in Action	78
Figure 7.1:	Bar Graph Voltmeter Circuit Diagram	82

Figure 7.2:	Using a Voltage Divider to Extend the Voltmeter's Range of Measurement	84
Figure 7.3:	Arduino Voltmeter Calibration Slope	86
Figure 7.4:	Voltage Divider with Trimmer Potentiometer for Adjustment	87
Figure 7.5:	Voltage Divider for Resistance Measurement	88
Figure 7.6:	Voltage Divider Measurement Function	88
Figure 7.7:	Picofaraduino	93
Figure 7.8:	Transistor Tester	93
Figure 7.9:	NTC Characteristic Curve	95
Figure 7.10:	Relative Output Voltage of an NTC with Voltage Divider	96
Figure 7.11:	NTC with voltage divider	96
Figure 7.12:	Temperature Output to the Serial Monitor	97
Figure 7.13:	AD22100 Temperature Sensor	98
Figure 7.14:	Temperature Measurement Using the AD22100	98
Figure 7.15:	Temperature Sensor with Ribbon Cable	100
Figure 7.16:	Digital Thermometer with 7-Segment Display	101
Figure 7.17:	Humidity Sensor Calibration	105
Figure 7.18:	Circuit Diagram for the Millamp-Hour Meter	107
Figure 7.19:	Discharge Graph of a 180 mAh Battery	108
Figure 7.20:	Proposed Battduino Construction in Enclosure	109
Figure 7.21:	Battduino Measurement Board	109
Figure 7.22:	Reflex Flasher	114
Figure 7.23:	Digital light meter	116
Figure 7.24:	Ultrasonic Transmitter Pulse Train	118
Figure 7.25:	Ultrasonic Transmitted and Received Pulses	119
Figure 7.26:	Ultrasonic Transmitter	119
Figure 7.27:	Ultrasonic Transmitter Circuit	119
Figure 7.28:	Ultrasonic Receiver	120
Figure 7.29:	Ultrasonic Receiver Circuit	120
Figure 8.1:	Graphical Displays for Toothbrush Timer	123
Figure 8.2:	Active Toothbrush Timer	124
Figure 8.3:	Toothbrush Timer Schematic	125
Figure 8.4:	Clock with LED Display	131
Figure 8.5:	Reaction Timer	136
Figure 8.7:	Transistor Driver Stage	138
Figure 8.6:	Universal Timer	139
Figure 8.7:	Transistor Driver Stage	142
Figure 8.8:	Part of the DCF77 Signal	143
Figure 8.9:	Circuit Diagram for DCF77 Module with LED and Transistor Stage	144

LIST OF FIGURES

Figure 8.10:	DSF77 Test Rig with LED and Transistor Stage	144
Figure 9.1:	Basic Structure of the I²C Bus	150
Figure 9.2:	I²C Message Bit Sequence	150
Figure 9.3:	I²C Timing	151
Figure 9.4:	PCF8574 Port Expander	152
Figure 9.5:	Schematic of a Mega Chaser with Three PCF8574 Port Expanders	153
Figure 9.6:	Construction of the Mega Chaser	153
Figure 9.7:	Schematic for Hexadecimal Debugger Display	156
Figure 9.8:	Hex Debugger Display Construction Example	156
Figure 9.9:	LCD via I²C on the Arduino	159
Figure 9.10:	I²C Temperature Sensors on the Arduino. Note the polarity – dimple on the right!	160
Figure 9.11:	LM75 for Local Measurements	161
Figure 9.12:	For Use as a Remote Thermo-meter, a Longer Cable May Be Attached to the LM75	162
Figure 9.13:	PCF8583 Real-Time Clock Module on Arduino	164
Figure 9.14:	Test Sketch for IR Remote Controls	168
Figure 9.15:	TSOP1736 IR Receiver Pin Assignment	168
Figure 9.16:	TSOP1736 IR Receiver on Arduino	169
Figure 9.17:	Decoding IR Signals Using the IRrecvDump Example Sketch	170
Figure 9.18:	IR-Controlled RGB Lamp	172
Figure 9.19:	IR-Controlled Digital Clock	175
Figure 9.20:	PS/2 Pin Assignment (Male Connector)	178
Figure 9.21:	PS/2 Mouse on Arduino	179
Figure 9.22:	Analysis of mouse signals in Serial Monitor	180
Figure 10.1:	Simple Alarm Tone Generator	186
Figure 10.2:	Circuit Diagram for an LM386-Based Audio Amplifier	187
Figure 10.3:	Construction Example for the Audio Amplifier	188
Figure 10.4:	A Suitable DIY Speaker Enclosure	188
Figure 10.5:	Sine, Sawtooth and Square Waves of the Same Frequency	191
Figure 10.6:	Theremin Circuit Diagram	196
Figure 10.7:	Theremin Construction	196
Figure 10.8:	Input Circuit for Reverb Generator	201
Figure 10.9:	Circular Buffer	201
Figure 10.10:	Synthesizer	204
Figure 10.11:	Synthesizer Construction Example	205
Figure 11.1:	Control Loop	208
Figure 11.2:	Lamp with Power LED	211
Figure 11.3:	Brightness Control Circuit	212

Figure 11.4:	'Gravity Compensation' Control Circuit	215
Figure 11.5:	Gravity Compensation Distance Ranges	216
Figure 11.6:	Levitation Distance as a Function of Coil Current	217
Figure 11.7:	The gravity compensator in action	217
Figure 12.1:	Servo Signal	222
Figure 12.2:	Model Servo Motors	222
Figure 12.3:	Photino	226
Figure 12.4:	Transistor Stage for the Electromagnet	228
Figure 12.5:	Cranino	229
Figure 13.1:	The Processing Interface	231
Figure 13.2:	Analog Monitoring	233
Figure 13.3:	Output of Analog Data to Serial Monitor	234
Figure 13.4:	Test Rig for All 6 Analog Channels	237
Figure 13.5:	Data Logger	238
Figure 14.1:	The Living Room Box	241
Figure 14.2:	Display Showing Time, Indoor and Outdoor Temperature, as well as Battery Voltage.	242
Figure 14.3:	Living Room Box Internals	242
Figure 14.4:	Living Room Box Block Diagram	245

List of Tables

Table 1:	Arduino ATmega168P and ATmega328 Microcontroller Features	16
Table 2:	The Most Important Variable Types	31
Table 3:	Frequently-Used Resistor Values	47
Table 4:	Important Capacitor Values	47
Table 5:	Arduino-to-7-Segment-Display Pin Assignments	61
Table 6:	Arduino-to-4x7-segment display pin assignments	63
Table 7:	Pin Assignment for Connecting Arduino to Dot-Matrix Display	67
Table 8:	HD44780-Compatible LCD display – Pin Assignment	76
Table 9:	Arduino-to-LCD Connections	77
Table 10:	Comparison between Reference and Measured Values	86
Table 11:	BPW40 Phototransistor Characteristics	113
Table 12:	Typical Illumination Examples	113
Table 13:	Structure of the DCF77 Signal	143
Table 14:	Decimal and Hexadecimal Numbers	155
Table 15:	Control of the RGB Lamp	172
Table 16:	Key Assignments for the Digital Clock with IR Remote	174
Table 17:	Servo Signal Timing	222

Index

2-D camera swivel 225
7-segment displays 62

A
Amplifier 185
Analog channels 233
Arduino 9
Arithmetic operators 31
Arrays 36
ATMega168 16
ATMega328 16
Audio processing 197

B
Bandpass filter 200
Bar graph display 59
Bar graph voltmeter 81
Bar graph 233
Batteries 107
Blink 24
Breadboards 43

C
Calibration 86
Circular memory buffer 200
Constants 35
Control loop 207
Control 215
Cranino 228
Crystal 123

D
Data loggers 237
DCF77 module 145
DCF77 radio module 123
Digital clock 130
Diodes 49
Displays 59
Dot matrix displays 67

E
Electrical appliances 141
Electrolyte 94
Electrolytic capacitors 89
Electromagnet 228
External voltage 15

F
Fuse bits 9

G
Geocaching 113
Gravitation compensator 214

H
Hexadecimal debugger 155
Humidity sensors 105

I
I2C bus 149
IDE 17
Illumination control 210
Inter Integrated Circuit Bus 149
IR receiver 171
IR remote control 174

L
LC displays 76
Levitation 217
Libraries 41
Light barriers 112
Light meter 115
Living room box 241
Living room hygrometer 105
Logical operations 32

M
Magnet 216
Microcontroller 15
Minimal Arduino 45
Mood lights 11
Multicolor LEDs 48

N
NiMH batteries 107
NTC 95

O
Optical sensors 112

P
PC keyboards 178
PCF8583 164
PD controller 218
Photodiodes 112
Phototransistors 112
Physical Computing 221
Potentiometers 47
POV 54
Power LED 53, 212

Power relay 141
Power supply 15
Printed circuit boards 43
Processing 231
Program icons 231
Prototyping boards 44
PS/2 mice 178
Pulse-width modulation 188
Pushbuttons 49

R
Random numbers 40
RC5 code 167
Reaction timer 135
Relative humidity 105
Resistors 46
Reverberation springs, plates 200
Reverberation 200
RGB LED 171
RTC 163
Running light effects 51

S
Sensors 81
Seven segment displays 62
Shields 13
Shift operators 33
Structures 36
Switches 49
Synthesizer 185

T
Temperature sensor 98
Theremin 195
Thermometer 95, 100
Timer 138
Toothbrush timer 123
Transistor parameters 93
Twilight switch 112

U
Ubuntu 23
Ultrasonic transducers 117
Upload 21
USB cable 20

V
Variables 30
Variable types 31
VCO 198
Verify 21

260